MOTORBOOKS
PowerTech Series ™

How to Tune & Modify

CHEVROLET
FUEL INJECTION

BEN WATSON

First published in 1997 by
Motorbooks International Publishers
& Wholesalers, 729 Prospect
Avenue, PO Box 1, Osceola, WI
54020-0001

Motorbooks International is a
certified trademark, registered with
the United States Patent Office

The information in this book is true
and complete to the best of our
knowledge. All recommendations
are made without any guarantee on
the part of the author or Publisher,
who also disclaim any liability
incurred in connection with the use
of this data or specific details

We recognize that some words,
model names and designations, for
example, mentioned herein are the
property of the trademark holder.
We use them for identification
purposes only. This is not an official
publication

Motorbooks International books are
also available at discounts in bulk
quantity for industrial or sales-
promotional use. For details write to
Special Sales Manager at the
Publisher's address

Printed in United States of America

Library of Congress Cataloging-in-
Publication Data

Watson, Ben
  How to tune & modify Chevrolet
fuel injection / Ben Watson. —
2nd ed.
    p.   cm. — (PowerTech series)
Includes index.
  ISBN 0-7603-0422-X
  (pbk. : alk. paper)
  I. Title. II. Title: How to tune
and modify Chevrolet fuel
injection. III. Series: Motorbooks
International PowerTech series.
TL214.F78W3697 1997 97-8466
629.25'3—DC21

On the front cover: New Chevrolet
engines no longer use carburetors to
feed fuel and air to the cylinders.
The modern fuel injection system,
consisting of an air plenum, a
control module, injectors, and a host
of other components, provides
dependability, great performance,
and low emissions. *David Gooley*

CONTENTS

INTRODUCTION
THE PERFORMANCE REVIVAL 4

CHAPTER ONE
**THE HISTORY
OF FUEL INJECTION** 6

CHAPTER TWO
**THE BASICS
OF ELECTRONICS** 12

CHAPTER THREE
TOOLS AND EQUIPMENT 28

CHAPTER FOUR
**HOW CHEVROLET
FUEL INJECTION WORKS** 34

CHAPTER FIVE
**AUTOMOTIVE
EMISSIONS AND THE LAW** 54

CHAPTER SIX
FUEL INJECTION TUNING 60

CHAPTER SEVEN
**FUEL INJECTION
TROUBLESHOOTING** 76

CHAPTER EIGHT
**ONBOARD DIAGNOSTIC
TROUBLE CODES** 90

CHAPTER NINE
HOW TO USE FAULT CODES 108

CHAPTER TEN
**TROUBLESHOOTING
FAULT CODES** 118

CHAPTER ELEVEN
**FUEL INJECTION PERFORMANCE
MODIFICATIONS** 140

CHAPTER TWELVE
**ENGINE MODIFICATIONS
AND THE LAW** 154

CHAPTER THIRTEEN
**AFTERMARKET FUEL-INJECTION
SYSTEMS** 156

CHAPTER FOURTEEN
**ON-BOARD DIAGNOSTIC SYSTEM
GENERATION 2** 158

APPENDIX 172

SOURCES 190

INDEX 191

THE PERFORMANCE REVIVAL

Good news! After years of lower and lower performance, after years of gutless econoboxes, after years of entering freeway traffic with white knuckles and closed eyes, power has once again returned to the American-built car. Unlike the classic muscle cars of the 1960s that brutally created power by sucking huge quantities of air and fuel into a quarter ton of iron, today's cars are technological marvels that coax and finesse every ounce of power from the air and fuel that they inhale.

Beginning in the 1960s and through the 1970s, issues of the environment and air pollution were of more concern to the automobile manufacturers than issues of good performance and driveability. It has been suggested that the manufacturers attempted to bluff or "politic" their way around the strict emission requirements imposed for the mid-1970s. The contention is that the American manufacturers did not do the same groundwork of research as their foreign competitors. Whatever the cause, the result was a return to the performance levels and driveability of much earlier decades.

As the 1970s progressed, compression ratios dropped to the level of the 1940s. Performance went out the window and with it went fuel economy.

Further complicating the situation in the late 1970s was the Arab oil embargo situation. At the same time that automotive air pollution was a major issue, America faced for the first time the realization that

cheap and plentiful gasoline was not a constitutional guarantee. Corporate Average Fuel Economy (CAFE) fines were levied against manufacturers building gas guzzlers. The manufacturers had two diverse sets of government-mandated criteria to meet: pollution and economy. Stop-gap systems were sold to the public in spite of known substandard driveability. The emphasis turned from delivering a quality product to just meeting Environment Protection Agency (EPA) and CAFE criteria.

All of this left little room for the backyard mechanic and even drove many professionals into another line of work. The 1960's tradition of buying a relatively new car and personalizing it for performance with fuel, intake, and exhaust modifications was relegated to nostalgia.

However, the manufacturers entered the 1990s with new generations of engines and fuel systems that addressed the problems of economy, emissions, performance, and driveability by reinventing the internal-combustion engine. Some of the new engines are based on designs that date to the era of the Beach Boys and "My 409," but changes in fuel systems, cylinder heads, and the tightening of production standards have brought these engines up to the level of 1990's technology. Others boast entirely new designs. The typical street car of the 1990s uses engine technologies that would have been seen only on the racetrack a few years ago.

Hot rodding is even beginning to return to popularity. Back in the sixties, hot rodding consisted of finding better ways of force-feeding air through the engine. Camshafts featuring greater lift and greater duration, freer-flowing intake and exhaust systems, carburetors big enough to stand up in, and final-drive ratios on a par with an aircraft tug all served to make for fun and speed. Those who were a little more affluent found ways of combining the freer flow of the engines with the ability to squeeze the incoming air tighter and tighter in the combustion chamber. Machined heads and pop-up pistons took compression ratios to the point where, on today's gas, these engines would ping while shut off in the driveway.

Many of these sixties' techniques are valid today for increasing the power of an engine, but there are new considerations on the scene. Fuel quality, government regulation, and emission controls mean that good planning of performance modifications on a street car become paramount—no longer can you just bolt on a bigger carburetor to solve a problem created by poorly planned or inefficient performance modifying.

Performance means a lot more than just running a quarter mile in less than 12 seconds or getting 25 miles per gallon. Performance also means dependability and reasonable maintenance costs, and this is where late-model cars really excel. Today's

systems are as easy to work on and maintain as any of the cars of the 1960s. Granted, on many of them you will need an arm that bends in four places just to change spark plugs, but the 429-ci Mustang was no different!

In spite of this, there is a fear—even among highly experienced professional mechanics—of working on late-model fuel-injected cars. A mystique has been associated with the onboard computer system; it is looked upon as a mystical concept that mere mortals are incapable of understanding, troubleshooting, or repairing. It seems odd to me that a generation which has progressed from Pong to destroying full-color Ninja mutants finds mystery in a box that merely reads voltages and turns transistors on and off.

The purpose of this book is to familiarize a lost generation of backyard, shade-tree, and weekend wrenches with the joys of maintaining and tuning their late-model fuelie. I also hope to educate the neophyte on the realities of high-tech fuel and ignition systems to avoid being ripped-off by unscrupulous or under-informed professionals. This book is as much a handbook for the consumer that never had and never will have grease under his or her fingernails as it is for the hands-on enthusiast.

Dig in and have fun!

CONTENTS

AVIATION
ADVANCES   7

HILBORN
INJECTION   7

CHEVROLET RAMJET
INJECTION   7

BENDIX
ELECTROJECTOR   7

BOSCH D-JETRONIC   7

CADILLAC-BENDIX
INJECTION   8

MODERN GENERAL
MOTORS INJECTION 10

Power in the 1960s meant bigger engines and bigger carburetors—nothing was left to finesse. This mid-'60s Corvette is typical of the days when performance meant a Rat motor and a Holley carburetor big enough to put your head through.

In 1957, Chevrolet introduced the American public to mass-produced fuel-injected engines with the "Fuelie" 283. The system, which was available on the Corvette as well as Chevrolet passenger cars, was vacuum-controlled and worked remarkably well when kept in proper "tune." This is a later-revision of that early system, on a 1963 Corvette. *Jason Scott*

CHAPTER ONE

THE HISTORY OF FUEL INJECTION

The history of fuel injection begins in the nineteenth century. Both N.A. Otto and J.J.E. Lenoir displayed internal-combustion engines at the 1867 Paris World's Fair. In 1875, Wilhelm Maybach of Deutz first converted a natural-gas engine to run on gasoline. This engine used a carburetor that featured a wick suspended across the flow of incoming air. The ends of the wick were submerged in gasoline held in a fuel bowl below the wick. When the engine was started the incoming air would pass across the wick, evaporate the gasoline and carry the fuel vapors up into the engine to be burned.

By the turn of the century Maybach, Carl Benz and their colleagues had added to one another's technology to the point where a dependable, float-level-controlled spray-jet carburetor had been developed.

As early as 1883, alongside the development of the carburetor, others were experimenting with crude fuel injection. Edward Butler, Deutz, and others developed the pioneering fuel injection systems.

Aviation Advances

Early advancements in gasoline fuel injection really came through aviation. From the very beginning, fuel injection played a major role in the development of practical aviation. In 1903 the Wright Flier used a 28-hp fuel-injected engine. Throughout Europe, prior to World War I, the aviation industry saw the obvious advantages that fuel injection afforded. Carburetors on aircraft are prone to icing during altitude changes, limiting available power; fuel injection is not. Carburetor float bowls are prone to spillage and fires during anything other than normal, level, controlled flight; fuel injection is not. However, World War I brought with it an emphasis on expediency and development costs; therefore, carburetor development pressed on while fuel injection was placed on a back burner.

The postwar prosperity of the 1920s saw a renewed but mild interest in the development of fuel injection. In the mid-1920s, Stromberg introduced a floatless carburetor for aircraft applications that is the predecessor of today's throttle body injection systems.

The military build-up which began in pre-Nazi Germany brought the Robert Bosch Company into the development of gasoline fuel injection for aviation. These early Bosch injection systems feature direct injection. Direct injection sprays the fuel under high pressure directly into the combustion chamber in much the same way as in a diesel injection system. In fact, the injection pump Bosch used for these systems was a modified diesel injection pump. During World War II fuel injection dominated the skies in all theaters of war. Late in the war, Continental used a fuel injection system designed by the SU Carburetter company of England and built in the United States by Simmonds Aerocessories on the air-cooled engine it developed for use in the Patton tank.

Electronic fuel injection found its beginnings in Italy when in 1940 Ottavio Fuscaldo became the first to incorporate an electrical solenoid as a means of controlling fuel flow into the engine. This began the path toward modern electronic fuel injection systems.

Hilborn Injection

After World War II, fuel injection development stopped. With the research and development money in the aircraft industry shifted away from internal-combustion fuel injection and toward jet engines, wartime improvements seemed destined for oblivion. The auto manufacturers were quite content with researching minor improvements to the dependable and inexpensive carburetor. Then, in 1949, a car equipped with a fuel-injected Offenhauser was entered at the Indianapolis 500. The injection system was designed by Stuart Hilborn and featured indirect injection. With indirect injection the fuel is injected into the intake manifold just ahead of the intake valve. If you were to compare this system to today's fuel injection, it could be said it was like having a throttle body injection system for each cylinder. It could also be compared to Bosch's CIS system (used in the VW Rabbit, Audi 5000, Volvo, and others) in that the fuel was not pulsed into the intake port but rather sprayed continuously, which earned it the nickname "constant flow" injection.

Chevrolet Ramjet Injection

In 1957, Chevrolet introduced its first fuel-injected engine for mass production in the Corvette. Borrowing heavily from the Hilborn design, the Rochester Ramjet fuel injection system was used by Chevrolet in 1957 and 1958, as well as by Pontiac in the 1957 Bonneville. The Ramjet system made use of a high-pressure pump that moved fuel from the tank to the injectors. The fuel was sprayed continuously ahead of the intake valve. A main-control diaphragm monitored the intake manifold pressure and engine load. This diaphragm was in turn connected to a lever that controlled the position of a plunger. A change in the position of the plunger-operated valve altered the amount of fuel diverted back to the pump reservoir and away from the injectors. This modified the air-fuel ratio to meet the needs of the engine.

This system suffered from a lack of understanding by those confronted with the day-to-day maintenance of the system. As a result, both Chevrolet and Pontiac dropped the Ramjet from their list of options in 1959. Chevrolet used it again as a Corvette option in the mid-1960s.

Bendix Electrojector

Also under development during the same time frame as the Ramjet system was the grandfather of the electronic fuel injection system destined for mass production. Design work began for this system in 1952 at the Eclipse Machine Division of Bendix Corporation, and in 1961 a blanket patent was issued on the Bendix Electrojector system. Almost simultaneous with the issue of the patent, electronic fuel injection was declared a dead-end project by Bendix management and was shelved.

Even though the Electrojector system itself never made it into mass production, it is the immediate ancestor of virtually all modern fuel injection systems.

Bosch D-Jetronic

When Bendix shelved electronic fuel injection in 1961, interest waned until 1966 when Bendix granted a patent license to Bosch. In 1968, Volkswagen introduced the Bosch D-Jetronic system into the US market on its Type 3 models.

The D-Jetronic system was used on a variety of European applications including Saab, Volvo, and Mercedes through the early 1970s until 1976. In spite of a lack of proper understanding of how the system worked by those confronted with servicing it, the system persisted and introduced the service and diagnostic procedures of electronic fuel injection to the American mechanic. And yet, this system was largely seen by the auto repair industry as a fluke, a one-shot deal.

Cadillac-Bendix Injection

Cadillac introduced the first mass-produced domestic electronic fuel injection system in September 1975 as standard equipment on the 1976 model Cadillac Seville. This system was developed through cooperation

Chevrolet's ultimate performance fuel package is the Tuned Port Injection (TPI) system found on the 5.0- and 5.7-liter Camaro and the 5.7-liter Corvette. Introduced in 1985, the TPI system offered a considerable power improvement over the preceding year's Crossfire Injection.

Chevrolet's mass-production venture into fuel injection in the 1980s was with Throttle Body Injection (TBI). Shown here is the single unit TBI assembly used on the 1982 2.5-liter Citation. Simplicity of design still makes TBI a popular fuel system on commuter cars.

For performance enthusiasts, Chevrolet's Corvette and Camaro featured the twin throttle body injection system that Chevrolet dubbed Cross-Fire Injection and which debuted in 1982. This exploded view clearly shows the two single-venturi throttle body units mounted on a common manifold base; note how the air/fuel charge from the left unit crossed the engine through the intake runners to feed the right bank, and vice versa. The relatively long intake runners resulted in increased torque. *Jason Scott*

Although mechanical fuel injection saw limited use on 1960s American cars, the Europeans, especially Mercedes-Benz, got involved quite heavily. This is a Bosch mechanical system on a mid-'60s six-cylinder Mercedes. The technology seen here is a direct descendant of Bosch diesel injection systems.

A closer look at the TPI system reveals the technology's leading edge at its introduction in 1985: the Bosch MAF sensor (ahead of accordian air duct) to monitor the amount of air entering the engine. In spite of its sophistication, the bulk of the TPI system can be tested with some basic tools such as the fuel-pressure gauge and the digital voltmeter, shown here.

Not all TPI systems use a MAF sensor, however. 1991–92 Camaros equipped with either the 5.0- or 5.7-liter TPI engines relied upon a MAP sensor to determine the volume and other characteristics of air entering the engine. These systems became known as "Speed/Density" systems because of the way they monitored air ingestion. *Jason Scott*

between Bendix, Bosch, and GM, and bore an amazing resemblance to the Bosch D-Jetronic system. By this time the manufacturers had begun to address the need for a systematized troubleshooting system to aid in the servicing and repair of fuel injection. These procedures have evolved into the flow charts that are the industry standard today.

The Cadillac-Bendix system was used until the introduction of the next technological improvement in fuel injection—the digital computer. Cadillac introduced its Digital Fuel Injection (DFI) system in 1980. The DFI system had been conceived as a multipoint system with one injector per cylinder. Simplicity and economy won out over technology, and DFI was introduced as a two-injector throttle-body injection system.

For Bendix, the idea for digital control of fuel injection dated back to patents it filed for in 1970, 1971, and 1973. Benefits that could be found with the digital computer included the more accurate control of the injectors, plus the ability of the computer to control a wide variety of engine support systems. With the use of a digital computer, a single compact control module could control ignition timing, a vehicle's air pump operations, torque converter clutch functions, as well as a wide variety of emission-related items. Additionally, the stored memory potential of a digital computer meant that it would be possible for the sensors to reprogram the computer for changes in the overall condition of the engine. This meant that the fuel injection system could detect and compensate for old, tired engines. Service intervals could be greatly increased because the injection system could compensate for the deterioration of ignition components. A digital computer could also detect and store memories concerning circuit failures in the system. These memories could later be called on by a technician to assist in troubleshooting.

Modern General Motors Injection

Pontiac was the first GM division to bring fuel injection to the masses. In 1982 it mounted a single-injector version of the throttle body injector system first used by Cadillac on its 2.5 liter Iron Duke. Also during that year, Chevrolet installed the TBI "Iron Duke" in its X-body Citation and F-body Camaro.

Chevrolet introduced the 5.0-liter Camaro and 5.7-liter Corvette engines with what was dubbed Crossfire Injection in 1982. This system featured two single-injector throttle body units mounted on a common manifold, with the left unit feeding the right side of the engine and the right unit feeding the left side of the engine. This crossfire concept allowed for increased air velocity and good fuel atomization. The Crossfire Injection System (CIS) provided a 20hp gain over carburetion on the 5.0-liter Camaro.

The Rochester Throttle Body Injection (TBI) system was also fitted to the Brazilian-built 1.8-liter overhead-cam engine in 1982, used in the J-body Cavalier. In 1983, its Pontiac cousin, the J-2000, was the first GM car to break the EPA 50 mpg barrier.

Although other GM divisions were using multipoint injection systems as early as 1984, Chevrolet held out until 1985 when it introduced the Port Fuel Injection (PFI) system on the 2.8-liter V-6 used in the Camaro, Cavalier, Celebrity, and Citation. Chevrolet's ultimate production fuel package was also introduced in 1985 on the 5.0-liter Camaro and the 5.7-liter Corvette engines; Chevrolet called the system Tuned Port Injection (TPI). TPI boasted an independently documented 434 cubic feet per minute (cfm) flow rate. To the old carburetor guys out there this may sound like a trifle when compared to an 850cfm Holley "Double Pumper," but a little math shows that the cfm potential of the 5.7-liter at 5,000 rpm is only 506 cfm. (Here we assume an impossible 100 percent volumetric efficiency. A more realistic cfm rating would be 405.) An entire section of

In 1992, Chevrolet introduced its "Generation 2" small-block Chevrolet V-8 engine, which included a new version of Tuned Port Injection. The 5.7-liter LT1 engine used a one-piece, short-runner intake manifold that produced more power at higher engine rpm than the long-runner, multi-piece TPI system used previously. The one-piece design also minimized the risk of air leaks, compared to the old design. *Jason Scott*

Under the Corvette's beauty covers lies a system that—intake configuration aside—functions nearly identically to the TPI system originally introduced in 1985. Note the fuel rails and injectors in their traditional locations. *Jason Scott*

this book is devoted to increasing potential cfm of not only the Port and Tuned Port systems but the Throttle Body systems as well. One well-documented report shows an increase in cfm up to 585 with only minor modifications to the intake system.

There is a lot of research and activity still going on with fuel

injection systems. Several aftermarket manufacturers are building and marketing throttle body and Tuned Port clones for older carbureted engines. The OEM (original equipment manufacturer) suppliers are looking again at the pluses and minuses of direct injection. As happens with many technologies, we

For 1996, Chevrolet replaced the aging Throttle Body Injection system on most of its trucks with a new fuel injection system—Sequential Central Port Injection (SCPI). A hybrid of TPI and TBI, the system uses a somewhat traditional throttle body/manifold configuration, but has an individual fuel injector for each cylinder, like TPI. The system relies on a Mass Air Flow meter to determine the volume of air ingested by the "Vortec" V-6 and V-8 engines. *Chevrolet Motor Division*

The 1997 Corvette also debuted with a new fuel injection system. Despite many subtle differences—a single-plate throttle body, a plastic intake manifold, and OBD-2 diagnostics—the system is functionally very much like TPI systems. Note the Mass Air Flow meter at the forward end of the air duct, at the air cleaner housing. *Jason Scott*

have reached a level of development and sophistication where simplicity and refinement are becoming the order of the day. Gone are the days of scrambling to simply produce a workable fuel injection system, for which much of the inspiration came from government-mandated emission and fuel-economy standards. The days of performance and a little fun are back once again.

In the years leading up to the publication of the first edition of this book, technology was changing very rapidly. Only five years earlier, the standard fuel system of the industry was the carburetor. Mechanics working in dealerships at that time were faced with almost daily changes in fuel systems. Additionally, the engineers were still developing fuel injection systems. This meant that there were still mistakes being made. Some of the most serious driveability problems of that time were a result of design problems, system deterioration, and programming mistakes. Today, fuel injection systems are far more dependable than carburetors ever were.

Since the first edition of this book, a lot of changes have occurred. The most important change is a slowing in the evolution of the technology. Along with that comes stability in technology. The typical dealership driveability technician used to groan whenever he heard about changes in the fuel system. There was a consensus of opinion that no change could be beneficial to the operation of the engine. Today, changes come in the form of fine-tuning and ease of troubleshooting and meeting the ever-changing demands of the Environmental Protection Agency.

At the core of these changes is a mandate that has become known as On-Board Diagnostics Generation 2 (OBD-2). I first heard about this mandate in the early 1990s. At that time, the rumors suggested that this technology would make diagnosing easier than ever. I am not sure these hopes have been realized. There is more about OBD-2 later in this book.

VOLTS	**12**
AMPS	**13**
WATTS	**14**
OHMS	**14**
OHM'S LAW	**15**
KIRCHOFF'S LAW	**15**
WAVEFORMS	**15**
MONITORED PARAMETERS	**17**
CONTROLLED FUNCTIONS	**17**
BASIC ELECTRONIC DEVICES	**18**
RADIO FREQUENCIES AND INDUCTION	**21**
ELEVEN CIRCUIT TYPES	**21**

C O N T E N T S

CHAPTER TWO

THE BASICS OF ELECTRONICS

Automotive electronics and the auto mechanic have been forced into an alliance that has been difficult for both parties. The auto mechanic, professional or amateur, relates to nuts, bolts, and steel. Automotive electronics is happiest in vibration-free, constant-temperature, and constant-humidity environments. The mating of these widely divergent personalities has resulted in much turmoil and frustration.

In the early 1980s, most professional technicians felt that electronics would disappear if they buried their heads in the sand long enough. I blame a lot of this attitude on members of my profession (automotive technician trainers) who either felt they needed to impress their students with the mystery of electronics, or lacked the background to simplify the subject to the barest of essentials necessary to effectively work with these systems.

In this chapter we will not address the topics of quantum mechanics, electron flow or unified field theory; on the contrary we'll stick to the fundamentals of automotive electronics, addressing only those topics which are necessary to understand, troubleshoot, and make repairs on modern GM fuel injection systems.

In order to work on these systems effectively, you will need to familiarize yourself with the terms described in this chapter.

Volts

A volt is a measurement of electrical pressure. Also known as electromotive force, it is often compared to pressure (psi) in a water system. The accurate measurement of voltage is critical to troubleshooting modern electronic fuel injection, since the computer gathers information about the functioning of the engine by measuring voltages and changes in voltage.

There are basically three ways that voltage is created in the automobile: electrochemically; by induction; and by static electricity.

Electrochemical Method

The most easily recognized example of the production of voltage through the electrochemical method is the car's battery. Here, chemical energy is converted to electrical energy by submerging two dissimilar metallic plates into a solution of sulfuric acid and water known as electrolyte. Although the primary job of the battery is to get the engine started, it also serves as both the basic and back-up power source for the entire electrical and electronic systems of the car. A problem within the battery can result in the Electronic Control Module (ECM) making errors in judgment that could result in driveability problems.

Another important electrochemical device in the modern automobile is the oxygen sensor. This device consists of a ceramic element made of zirconium dioxide that becomes conductive for oxygen ions when heated to about 600 degrees Fahrenheit. This element is shaped like a thimble. On both the outer and inner

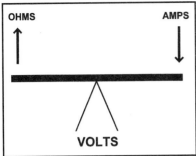

Although much discussion could be spent on Ohm's Law, the most critical concept is that the circuits possessing high resistance will have a small current flow. The various sensor circuits we will cover in this book fall into this high-resistance, low-current flow category. Many of these circuits are protected from excessive current flows by means of a current-limiting resistor in the ECM.

surfaces of the sensor there is a thin, gas-permeable layer of platinum. A channel down the center of the sensor allows outside air, with a constant 21 percent oxygen content, to contact the layer of platinum on the inside of the thimble. The layer of platinum on the outside is exposed to the exhaust gases. As the air-fuel ratio delivered to the engine varies, the oxygen content of the exhaust gasses varies. A difference in oxygen content on the two sides of the ceramic thimble will cause a voltage to be generated.

Induction Method

There is a good side and a bad side to the induction method of creating a voltage. When a conductor passes through a magnetic field, or when a magnetic field passes across a conductor, a voltage is generated in that conductor. The amount of voltage depends on the strength of the magnetic field at the point where the conductor passes through it, the size—both length and diameter—of the conductor, and the speed at which the conductor passes through the magnetic field.

There are two primary places in the automotive electrical system where the production of voltage through induction is done deliber-

ately. In the alternator, an AC (alternating current) voltage is induced into the stator windings. Since the automobile battery will not accept AC, this voltage is converted by means of a diode rectifier bridge to DC (direct current) before it leaves the alternator. The resulting voltage output is between 13.5 and 14.9 volts DC.

The second place where deliberate induction occurs is in the ignition coil. Here, a voltage of up to nearly 60,000 volts can be induced as the coil control transistor switches off the current flow through the primary windings of the coil, causing the magnetic field around the primary windings to collapse across the secondary windings. This voltage is then made available to the gap of the spark plugs.

There can also be a down side to induction. When a conductor carrying a relatively large current lies adjacent to a conductor carrying either a small current or no current at all, the potential for induction exists as the current is shut off. An induced voltage has the capability of activating a controlled device or injecting a false signal into a sensor circuit. When a wiring harness has been removed for maintenance, particularly the computer's wiring harness, care should be taken to ensure that it is returned to its original physical position away from sources of induction.

Static Electricity

There are no places on the automobile where static electricity is deliberately generated. Static electricity is not a new concern to the automotive world. I remember as a kid in Amarillo, Texas, in the wintertime touching the door handle of our old Buick and receiving quite a jolt. Many people had rubber "static straps" hanging from their rear bumper.

In those days, however, concerns about static were purely a matter of comfort. Today, static can spell death to the electronic control module. Those who live in static-prone parts of the country have grown to accept the little doorknob snaps as

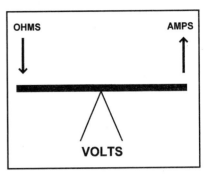

The driver, or output device, circuits have less resistance in them and therefore a greater current flow. These larger current flows are necessary to generate magnetic fields large enough to activate the solenoids and relays that carry out the operational functions of the engine. Corrosion and other sources of resistance can affect the strength of the magnetic fields and therefore the functioning of the output circuits.

simply a part of life. These little snaps represent a tremendous amount of voltage. Feeling one of these sparks represents a voltage discharge of over 1,000 volts; feeling and hearing it represents over 3,000 volts; and feeling, hearing, and seeing it represents over 5,000 volts! These voltages are well in excess of the 17 volts that the ECM can safely accept. When handling the ECM and other electronic modules, extreme care should be taken to limit the potential of damaging them by means of static discharge.

Risk from static discharge damaging electronic devices can be reduced by grounding yourself as you handle these devices. Your local electronics supply store, such as Radio Shack, can sell you a static strap for as little as $4 or $5.

Amps

An ampere is the flow rate of the electrical current. Like gallons per second in a pressurized water system, amperage is the volume of electricity passing a given point.

This is not a new measurement to the automotive technician, but historically he or she has been involved in troubleshooting circuits carrying anywhere from a quarter amp to several hundred. The electronic fuel injection system has circuits that carry as little as a milliamp (0.001 amp). When you are dealing with such small currents, using the wrong test equipment can lead you to believe that there is a fault in a circuit that has no problem at all.

Watts

A watt is a unit of power. Like horsepower, wattage is the amount of energy being expended by the electrical circuit in which it is being measured. As a measurement of power, watts are sometimes used in place of horsepower. There are 746 watts in 1 hp. Mathematically, watts equals volts times amps. A circuit flowing 2 amps at 12 volts is dissipating 24 watts of power.

Ohms

Ohms are not so much a measurement of electricity as it is a measurement of the quality of a conductor. An Ohm is the unit used to measure resistance of the conductor to current flow. In the past, ohmage has been an important factor in considering why a circuit did not work, and what we were looking for was a point in the problem circuit where the resistance was too high. For instance, if we were dealing with a starter that had a slow cranking speed, the possibilities would include a bad starter, a weak battery, or high resistance in the power or ground cables. For circuits like this—ones which almost every car owner has had exposure to—the tiniest bit of resistance will have a major effect on its operation.

For example, a typical cranking voltage is in the vicinity of 10 volts. For a small-block Chevrolet, a typical current flow through the starter circuit is around 200 amps. Just 0.01 ohms of resistance can cause a reduction in available voltage to the starter of 2 volts. This is a significant effect from a relatively insignificant amount of resistance. The reason the effect is so significant is because of the high current flow in the circuit.

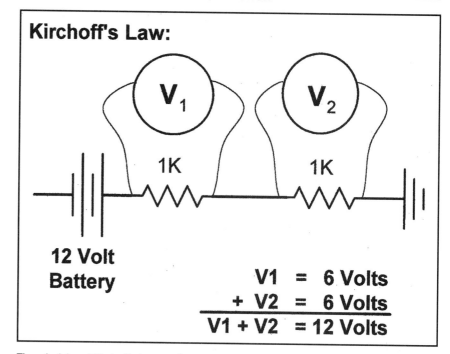

The principles of Kirchoff's Law are fundamental to the way the ECM gathers data from most of its sensors. As current flows through a series circuit, voltage is reduced. By the time the current reaches ground, all of the voltage is gone. This drop in voltage is proportional to the total amount of resistance in the circuit. For instance, if the total resistance of a circuit with a source voltage of 12 volts is 2,000 ohms, the voltage will drop 6 volts when it passes through 1,000 ohms.

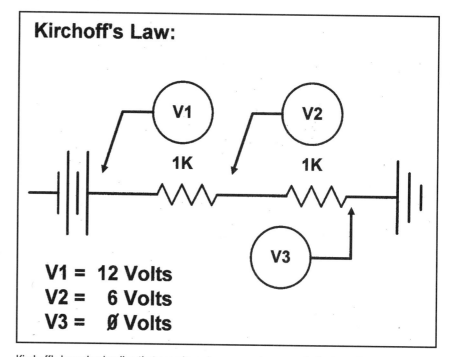

Kirchoff's Law also implies that as voltage is measured progressively through a series circuit, the voltage detected will be smaller and smaller. The ground of any series circuit should have 0 volts. If the voltage at the ground is higher than 0, then there must be resistance at the ground connection.

Sine Wave

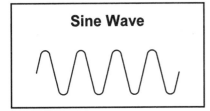

A sine wave results when a device such as a reluctance pickup coil produces a voltage that slowly builds, then slowly drops. In automotive applications, however, a sine wave is usually related to AC voltage, such as those produced by the alternator and the pickup coil. In reality, a sine wave can also be DC.

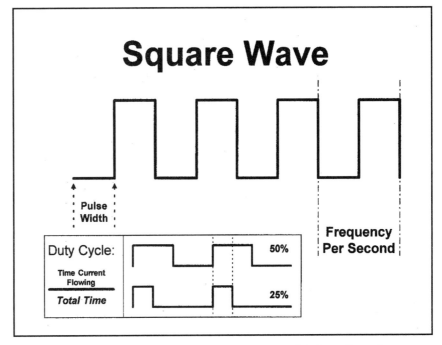

The characteristics of a square wave often need to be measured during a diagnostic procedure. The two that are most often measured are frequency and duty cycle. The frequency is the number of complete highs and lows occurring per second. Frequency can be measured with a digital tachometer. (Use the four-cylinder scale and divide the tach reading by 30.) Duty cycle is a percentage measurement of the relationship between the high and low of a pulse. This can be measured with a digital dwell meter set on the four-cylinder scale (multiply the reading by 1.1).

In the past, we have been concerned primarily with looking for high-resistance problems in the range of hundredths to a dozen or so ohms. Today, we might just as easily be looking for a problem in an electronic fuel injection circuit where the normal resistance is several hundred to several thousand ohms. In a practical sense, for some of these circuits, hundreds of ohms may make little or no difference in their functioning. For other circuits, such as those that measure temperature, resistance is critical and they are sensitive to a few dozen ohms.

Ohm's Law

If you ever get an opportunity to teach a vocational school class for mechanics who have been working in the field for several years, and if you like to hear really deep, heartfelt groans, bring up the subject of Ohm's Law! For our purposes, there is no need to go through lengthy mathematical equations. Let us just stick to what is relevant to our needs.

What Ohm's Law says is that there is a relationship between volts, amps, and ohms. When the resistance in a circuit changes, but we keep the voltage the same at the source, then the current flow will change. In other words: Increase the resistance (ohms) and the current flow (amps) will decrease; decrease the resistance and the current flow will increase. Decrease the resistance in the circuit

too much and you let the smoke out of it. Since these circuit computers do not gather data or control actuators through changes in current flow, for the real world of troubleshooting electronics in fuel-injected cars, Ohm's Law is not of much use—except if the resistance in the wiring harness of a circuit (particularly an actuator circuit) drops close to zero. Then you will probably damage the Electronic Control Module (ECM).

Kirchoff's Law

Everybody talks about George Simon Ohm, but really it is Gustav Robert Kirchoff who deserves a round of applause for helping us gather engine data and troubleshoot onboard computer systems. Kirchoff discovered these important factors: First, the sum of the voltage drops in a series circuit is always equal to source voltage; second, the algebraic sum of the current flowing toward a single point is zero.

We automotive computer diagnosticians do not have a lot of direct use for Kirchoff's Second Law, but we have a great deal of use for his First. A voltage drop occurs as a current flows through a resistor. Using our water system analogy, when a current of water being moved by pressure passes through a restriction, the water pressure on the downstream side of the restriction will be less than the pressure on the upstream side. As an electrical current flows through a resistor, the voltage (pressure) on the outbound side will be less than the voltage on the inbound side.

Voltage drop measurement not only allows the ECM to gather data on engine operating parameters, but it will also become a primary diagnostic tool.

Waveforms

When voltage changes in a regular or rhythmic fashion it is referred to as a waveform. The only way to really see these waveforms is with an oscilloscope. Automotive electronics deals with two types of waveform patterns: the sine wave and the square wave.

Sine Wave

Produced as a voltage, the sine wave slowly builds to a peak, then slowly decreases to a valley or trough. Components such as alternators and reluctance-type ignition pickup coils produce a sine wave. Seldom will you need to measure any part of the sine wave; noting its presence is sufficient. See the section in chapter 3 on AC voltmeters for more detail.

Square Wave

The square wave is a little more complex than the sine wave. A square wave is an on-off pulse created to monitor or control various functions. Being an on-off signal, the voltage does not build slowly and then decrease; the voltage measured is either high or low. This waveform is created by such things as the Mass Air Flow (MAF) sensor and the Hall Effects sensors to monitor, and by the ECM to control the opening and closing of injectors and the switching on and off of the ignition coil.

There are four characteristics of the square wave that you may need to measure: amplitude, frequency, duty cycle, and pulse width.

Amplitude

Amplitude is the amount of voltage change that occurs as the current flowing through a circuit is switched on and off. This is a measurement you will seldom if ever take. Amplitude measurements cannot be accurately taken without an oscilloscope.

Frequency

Frequency is the number of complete on-off cycles that occur in a given time frame. The most common unit of measurement for frequency is a hertz (Hz). A frequency of 10 Hz means that there are 10 complete "ons" and 10 complete "offs" occurring each second. Several Chevrolet fuel injection sensors produce a variable-frequency

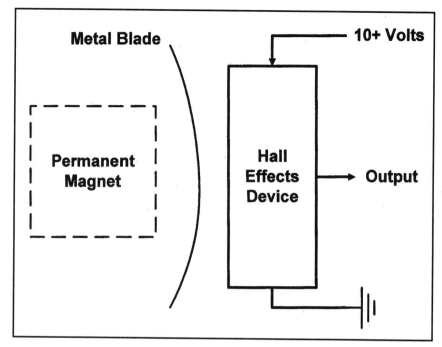

Among the sensors that produce a square wave is the Hall Effects sensor. The Hall Effects is used to measure the rotational speed and position of such things as the distributor, crankshaft or camshaft.

Monitored Parameters	Controlled Parameters
A/C Power Steering	Air Pump
Coolant Temperature	Fuel Timing
Engine Load	Idle Speed
Intake Air Flow	Knocking
Intake Air Temperature	Radiator Fan
Knocking	Timing
Load Exhaust Oxygen	Torque Converter Clutch
RPM	
ThrottlePosition	
Transmission Gear	
Vehicle Speed	

The Chevrolet computer controls a variety of functions by taking in data from the sensors that monitor engine operating conditions, analyzing it, then sending a signal to the controlling devices (or actuators). In general, the monitoring of sensing occurs at 5 volts while the controlling, or actuating, occurs at 12 volts.

square wave, including the crank sensors on some distributorless engines and the Delco-style MAF sensor.

Duty Cycle

Duty cycle is a difficult concept to fully appreciate. It is the measurement of the length of time current is flowing through the circuit and the time current is not flowing through the circuit, measured in percent. The concept is not new to the automobile engine. For decades we have measured the relationship between the

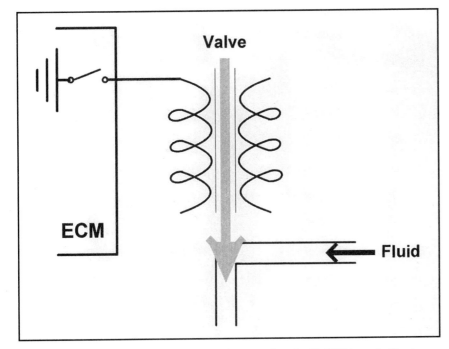

The normally-closed solenoid-operated valve does not allow the "fluid" it is controlling to flow when the solenoid is not energized. A solenoid like this might be used to control the flow of vacuum to the canister purge valve. Before the ECM grounds the solenoid there is no vacuum to the canister purge valve; therefore, the evaporative canister will not be allowed to purge. When the ECM grounds the solenoid coil, the valve is opened, allowing vacuum to flow to the purge valve, which in turn allows the canister to purge. Similar devices might be used to control vacuum to the EGR valve or fuel into the engine (injectors are a normally-closed solenoid-operated valve). Normally-open solenoid-operated valves allow fluid to flow when the solenoid is not grounded.

time the current is flowing through the ignition coil and the time that current is not. We called that relationship dwell. Dwell is the duty cycle of the primary ignition circuit measured in degrees of rotation.

Pulse Width

Pulse width is the length of time, measured in seconds, minutes, hours, or days, that an actuator is energized. For our purposes, the measurement of pulse width is almost exclusively reserved for injector on-time. For injector on-time, the unit of measurement is milliseconds (msec).

Chapter 3 covers techniques for measuring the four square wave characteristics with the simplest, most common tools possible.

Monitored Parameters

Monitored Parameter is a blanket term used for computer input. On late-model Chevrolet fuel injec-tion, this input would include the following:
- RPM
- Coolant temperature
- Air temperature (not on all applications)
- MAF (only on PFI and TPI applications)
- Manifold pressure (not on all applications)
- Barometric pressure (not on all applications)
- Exhaust oxygen
- EGR (exhaust gas recirculation) function
- Throttle position
- Others, depending on the application

The monitoring of the various functions around the car usually happens between 0 and 5 volts. Five volts is used as a reference voltage. When the ECM sees 5 volts from one of its sensors, it usually interprets that to mean whatever is being mon-

Resistor Symbol

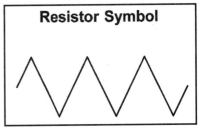

The resistor is one of the most basic electronic devices. Its primary task is to reduce or limit current flow; as a by-product of doing this, it also produces a voltage drop. Depending on the circuit design, a resistor might be there to perform either task. One of the most important jobs that the resistor has is to limit current flow to the various sensors and actuators in the fuel-injection system to protect the ECM's power supplies and voltage regulators.

Thermistor Symbol

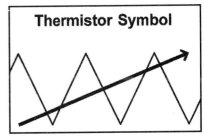

This symbol represents a thermistor. The thermistor is an electronic device whose resistance changes as the temperature is exposed to changes. The thermistor Chevrolet uses to measure coolant and air temperature is called a negative temperature coefficient (NTC) thermistor. With the NTC thermistor, as the temperature goes up, the resistance drops. At -40 degrees Fahrenheit this resistance is about 100,700 ohms; at 210 degrees Fahrenheit it's about 185 joules.

itored by that circuit is occurring to its maximum.

Controlled Functions

Most functions and operations around the car are controlled by the computer with a solenoid-operated valve. This valve can control vacuum, fuel, EGR, or airflow. There are two types of solenoid-operated valves to control these functions: Normally-open and normally-closed valves.

Normally-open valves allow whatever is being controlled (vacuum and so on) to flow through the valve when current is not flowing through the solenoid. Typically, a normally-open valve is used where either the device needs to receive vacuum (or whatever is being supplied) during most of the time the vehicle is in operation, or when the limp-home, or limp-in mode requires vacuum to be applied.

Normally-closed valves do not allow vacuum (fuel, air, or EGR) to flow when current is not flowing through the solenoid. The best example of this in the injector itself. When the solenoid does not have current flowing through it, the valve is closed, keeping fuel from flowing through the injector. The normally-closed valve is used when the normal state for a system during driving is to not receive vacuum, fuel or whatever. As odd as it may sound, the fuel injector is not usually open when driving. The duty cycle of the injector varies, depending on driving conditions, between 1 percent and 20 percent duty cycle, which means that although it is opening and closing continuously while driving the car, it is spending most of its time closed.

The normally-closed valve is also used where we want the limp-home mode for a given function not to allow the flow of vacuum, fuel, EGR, or air.

Solenoid-operated valves are not the only way that things are controlled. The ECM will control some things directly. Transistors or "quad drivers" supply the ground through the ECM for the Check Engine light, fuel pump relay, the air conditioner compressor clutch and the injectors. Some manufacturers also control the alternator field through the ECM.

It could also be said that the ignition timing is controlled directly by the ECM. Actually, the ECM sends a signal to the ignition module, which in turn controls the ground for the ignition coil.

The following functions are among those controlled by the ECM. Note, however, that not all of these functions will be controlled by the ECM in the car you are working on.

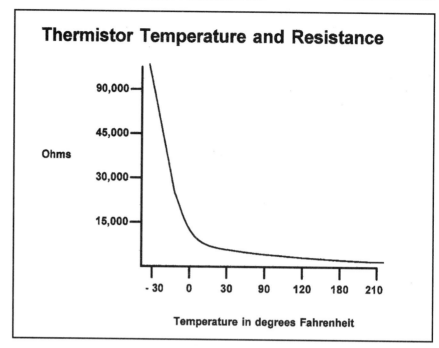

From -30 degrees to about 50 degrees Fahrenheit the resistance of the coolant and air charge temperature thermistors drops rapidly. From 50 degrees to over 200 degrees Fahrenheit the resistance change is much more gradual.

Solenoid control:
- Canister purge solenoid
- Air Management (air pump) System (AMS)
- EGR operation
- EGR position control
- Lock-up torque converter

Direct control:
- Check Engine (Service Engine Soon) lamp
- Fuel-pump relay
- Injectors
- Alternator field
- A/C compressor clutch
- Ignition timing
- Others

Basic Electronic Devices

It will be helpful as you progress through this book to become familiar with the electronic basis for many of the sensing devices that are used. Since this book is not intended to be a course in electronics, each item will be covered only with respect to the job it performs in a fuel injection system.

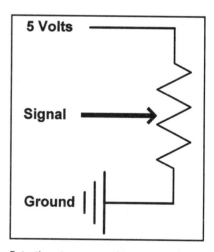

Potentiometers are used to measure the position of such things as the throttle or the 2.8- and 3.1-liter EGR valves. As the position of the monitored item changes, the voltage leaving the sensor on the signal wire will change. A typical closed position voltage would be about 0.5 volts, while the wide-open voltage exceeds 4.0 volts.

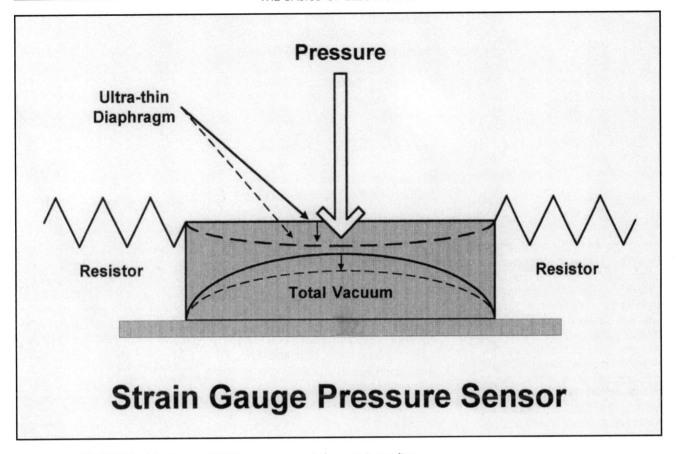

Pressure

Ultra-thin Diaphragm

Resistor

Resistor

Total Vacuum

Strain Gauge Pressure Sensor

The Chevrolet Manifold Absolute Pressure (MAP) sensor uses a strain gauge to monitor manifold pressure. The strain gauge consists of a thin, flexible silicone chip that acts as a diaphragm to stretch resistors located around the edges of the chip. As these resistors are stretched, their resistance changes. The end result is a pressure-sensitive device that behaves like a potentiometer to produce a voltage that changes with pressure changes.

Resistor

A resistor is an electronic device used to limit current flow and reduce voltage within a series circuit. Two types are common in automotive use today. The first is the wire-wound resistor which is commonly found in high-current uses such as the ballast resistor in the old point-condenser ignition system. The other type is the carbon resistor. The carbon type is used in low-current-flow circuits such as the ECM.

One of the most important uses of the resistor in the ECM is to limit current flow into and through the ECM. Most of the output or driver circuits of the ECM have a resistor in series with the output to prevent the driver circuit from overloading should the wire from the ECM to the actuator become grounded or shorted to voltage.

Thermistor

The thermistor is a resistor used to monitor temperature. The type that Chevrolet uses is called a negative temperature coefficient thermistor. As the temperature of what is being measured increases, the resistance drops. At -40 degrees Fahrenheit this device will have 100,700 ohms. As the temperature it is being exposed to increases, the resistance slowly drops to about 185 ohms at 210 degrees Fahrenheit.

Thermistors are used to measure the temperature of the coolant and intake air on almost all Chevrolet fuel injection applications. The Mass Air Flow sensor also uses a thermistor to measure the temperature of the air passing through it to assist in the measurement of the air mass.

Although potentiometers can be used to monitor a variety of functions around the engine, Chevrolet uses them mainly to measure throttle angle or position. This TPS (throttle position sensor) is an adjustable unit.

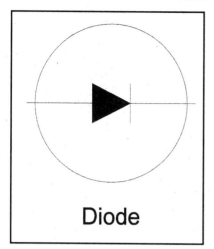

The diode is a one-way electronic valve. Under normal conditions the diode will allow current to flow in one direction but not in the other. This allows the diode network in the alternator to convert alternating current into direct current. Diodes are also used to prevent high-voltage spikes caused by the switching off of the magnetic field in the air conditioner compressor clutch from spiking the circuitry in the ECM. In this usage the diode is called a suppression or clamping diode.

Potentiometer

The potentiometer is a resistor that utilizes a metal wiper that moves back and forth across a carbon element. It is used to sense or detect the physical location of a moving device. A potentiometer has three connectors: one connected to the 5-volt reference, a second connected to ground through the ECM and a third connected to the input section of the ECM to detect the position of what is being measured as a varying voltage.

Potentiometers are used to measure throttle position and in some cases EGR position.

Strain Gauge

The strain gauge is the principle component in Chevrolet's Manifold Absolute Pressure (MAP) sensor. A strain gauge consists of a silicone chip approximately three millimeters (mm) square and 250 micrometers, or microns (1 micron equals 0.000001 meter) thick. The center of the square is only about 25 microns thick to form a diaphragm. The edges of the chip are sealed to a Pyrex plate with a vacuum between the chip and the plate. A set of four resistors around the edges of the plate form a Wheatstone Bridge; these resistors are sensitive to being stretched by the flexing of the silicone chip.

As manifold or barometric pressure on one side of the chip works against the vacuum on the other side, the resistors are stretched and contracted, weakening their resistance to change. The strain gauge MAP has a 5-volt reference, a ground and a wire that carries pressure information as a variable voltage to the ECM.

Diode

There are two important types of diodes to automotive electronics, the Light Emitting Diode (or LED) and the photodiode. In automotive language, when we speak of diodes we normally think of the devices in the alternator that convert AC voltage into DC. We call them a one-way electronic valve. For our purposes, it might be more appropriate to think of these diodes as a light bulb and a light-sensitive switch.

The LED we will most often refer to in this book produces infrared (invisible) light when current is flowing through it. Pair this with an infrared photodiode which allows current to flow when this type of light falls on it, and you have a sensor capable of detecting the presence of an opaque object between them.

Transistor

Dealing with automotive fuel injection systems, we really do not have to be concerned with the functioning of transistors or their testing. A transistor might be referred to as an electronic relay. Transistors are used as output or as controlled device drivers by the ECM. Some applications use an integrated circuit (IC) that contains four output transistors, known as quad drivers.

Transistors have three connectors: the base, the collector, and the emitter. Voltage applied to the base can be compared to voltage applied

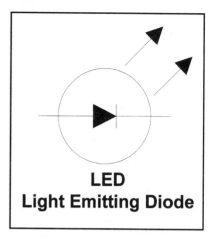

The LED or Light Emitting Diode is used as a light source for a wide range of functions. The most obvious is where the LED is used as a warning or indicator lamp. Less obvious is in the cable-driven speedometers where an LED (transmitter) and a photo diode (receiver) are teamed up in the back of the speedometer head to produce a signal directly proportional to vehicle speed.

to the pull-down winding of a standard electromechanical relay. The collector and emitter form the connections for the switch that connects power to the controlled device.

There are two types of driver transistors commonly used in automotive electronic systems. The first is called the NPN transistor, which will have the collector connected to the ground side of the controlled device; the other side of the controlled device will be connected to power. The emitter of the transistor is then connected to ground. When a high voltage (for example, 5 volts) is applied to the base of the NPN, the collector becomes connected to the emitter, thus grounding the controlled device.

The second type, the PNP transistor, will have the plus side of the controlled device connected to the collector, with the emitter connected to power. The controlled device will be grounded to the engine block or in back of the battery. When the base is grounded the PNP connects power from the emitter to the collector, which then applies power to the controlled device.

For most controlled devices in Chevrolet's electronic fuel injection system the NPN transistor is used. Chevrolet's preferred method of controlling the various actuators around the engine is by grounding them through the ECM. This makes the NPN the natural choice.

Microprocessor

At the heart of the Electronic Control Module (ECM) are three major types of microprocessors used for memory storage and decision making: the ROM, PROM, and RAM.

ROM

ROM stands for Read Only Memory. This microprocessor contains the basic program of the ECM. It is the part that says, "When I see this happen, I have to make that happen." The ROM features a non-volatile memory, which means that even when power is taken away from the ROM, it will retain its programming and memory indefinitely.

PROM

PROM, the Programmable Read Only Memory, is the fine-tuning or calibration microprocessor. Like the ROM, the PROM is also nonvolatile. This chip contains information about the specific car in which the ECM is installed. Types of information include the following:
• Vehicle size
• Weight class
• Wind drag
• Rolling resistance
• Engine size
• Final-drive ratio
• Type of transmission
• Camshaft design
• Emission control devices used

Information from the PROM is used by the ROM to assist the ROM in making decisions. When engine modifications are made on a late-model fuel-injected Chevrolet, the PROM must be replaced with one containing a high-performance program.

RAM

RAM, the Random Access Mem-ory, is used by the ECM for the temporary storage of information or to perform mathematical computations. Additionally, the ECM stores information here about the air-fuel history of the engine and faults that have been detected in the sensor and actuator circuits of the fuel injection system.

Radio Frequencies and Induction

The spark-ignition engine is an extremely hostile environment for transistor-based systems to operate around. There are radio frequencies generated wherever there is a spark jumping an air gap, such as in the distributor cap or across the spark plugs, which can interfere with the proper operation of transistors.

There are also sources of induced voltages such as secondary ignition and the alternator. These induced voltages can activate a controlled device at the wrong time. The NPN transistor circuit was chosen to drive most actuators because this circuit has the actuator powered all the time and grounded by the ECM. It is impossible for a ground to be induced in a circuit that is continuously powered; therefore, it is unlikely that the controlled device will be falsely energized.

Eleven Circuit Types

The good news about electronic fuel injection and engine-control systems is that there are really only 11 types of electronic circuits that are used. Each of the sensor and actuator circuits will fit into one of these categories. Familiarize yourself with each of these circuits. Then, working with and troubleshooting the circuits will be much easier. The first 9 are sensor circuits.

Switch to Voltage

The switch-to-voltage circuit involves a need to sense when an event occurs. As an example, let us say that we have an onboard computer system that needs to know when the driver is sitting in the seat. We would put a switch in the seat, one side of which would be con-

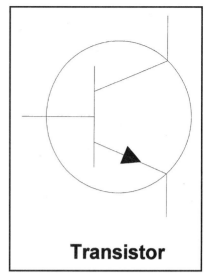

Transistor

The transistor plays many roles in the operation of the ECM. One of the most critical is its role in switching various actuators on and off. Acting like a semiconductor relay, the transistor drivers of the ECM respond to micro-processor commands to switch the injectors, air control, EGR, and many other solenoid- and relay-controlled functions.

nected to power at all times; the other would be connected to the computer. When the driver sits in the seat the switch would close, allowing power to be connected to the computer. The computer would know that the event has occurred.

This circuit assumes that the switch used is a normally open switch. Another circuit that works the same way might use a normally closed switch. If a normally closed switch is used then voltage will be applied to the computer until the event occurs. At that time the switch will open, causing the voltage at the computer to drop to 0 volts. The computer is then made aware that the event has occurred.

The switch-to-voltage circuit, though used extensively by other manufacturers, is seldom used by Chevrolet.

Switch to Pull Low

In the second type of circuit, the switch to pull low, the ECM applies a reference voltage (usually 5 volts) to one side of a switch through a current-limiting resistor. The other side of the switch is connected to ground. The computer monitors voltage on the circuit at the outbound side of the resistor. When the event occurs, the switch closes, grounding the outbound side of the current-limiting resistor and causing the voltage senses by the computer to drop to 0. The computer knows that the event has occurred.

As described, this circuit uses a normally-open switch. It could also use a normally-closed switch, however, so that when the event occurs, the voltage registered by the computer will rise. An example would be the monitoring of vacuum applied to the EGR valve.

Variable Resistance to Pull Low

The variable-resistance-to-pull-low circuit operates much like the circuit just described. In this circuit, however, the switch is replaced with a variable resistor, usually a thermistor. As an event occurs, such as an increase in engine temperature, the resistance in the resistor decreases, causing the voltage on the outbound side of the current-limiting resistor to decrease. The computer knows that the event is occurring. This decrease in voltage could be seen by connecting a digital voltmeter to the wire carrying the reference voltage to the computer.

Examples of this type of circuit include the coolant temperature sensor and the air charge temperature sensor.

Variable Resistance to Push Up

The only difference between the variable-resistance-to-push-up circuit and the variable-resistance-to-pull-low circuit is that when the event occurs, the resistance in the sensing resistor increases, causing the voltage seen by the computer at the outbound side of the current-limiting resistor to increase. The computer then knows that the event is occurring.

The knock sensor circuits, which do not use a separate ESC (Electronic Spark Control) module, are an example of this type of circuit.

The ECM uses microprocessors to gather information, process that information, and signal driver transistors to turn the various actuator circuits on and off. The three main components are RAM (temporary memory), ROM (basic computer program) and PROM (program fine-tuning).

Three Wire Variable Voltage

The three-wire-variable-voltage circuit has the computer supplying a constant reference voltage (again, usually 5 volts) and a ground to the sensor. A third wire carries information about the changing condition or position of a component from the sensor to the computer.

On the first four circuit types mentioned, the computer reads the information on a voltage line going out; on this type, the computer reads information coming back in.

Three examples of three-wire variable voltage circuits are the throttle position sensor, the MAP sensor, and the Bosch-style MAF on Tuned Port Injection applications.

DC Frequency Pulse Generator

The DC frequency pulse generator usually consists of some type of integrated circuitry that is used to monitor the condition of a component. Power (either 5 volts or battery voltage) and ground are fed into the sensor to power its circuitry. The sensor creates a square wave, the frequency of which varies as the condition being monitored varies. The voltage source of the square wave can either be the computer, which means that the sensor grounds and un-

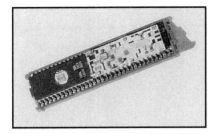

The PROM contains information about the size, weight, engine, final-drive ratio, transmission, cam design, and emission control devices. A factory authorized PROM change is sometimes the only effective cure for a drivability quirk.

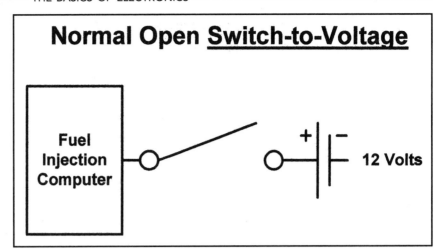

Of the 11 basic types of circuits used in electronic fuel-injection systems, the simplest sensor circuit involves the use of a simple switch that connects battery voltage to the ECM when an event occurs. An example of a circuit where this is used is the switch to voltage circuit.

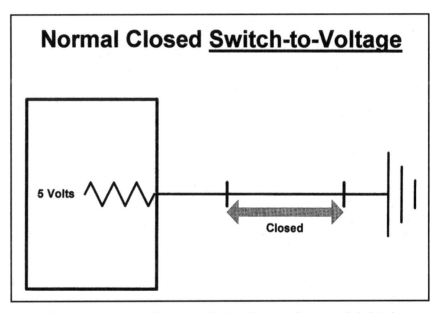

A second type of circuit uses either a normally-closed or normally-open switch that changes a reference voltage as the event controlling the switch occurs. In the circuit pictured, the switch is closed until the event, at which time the switch opens. Before the switch opens, the ECM's microprocessors are detecting 0 volts on the outbound side of the internal resistor. When the switch is opened by the event, the voltage rises, signaling the ECM. An example of this circuit is the 2.8-liter Generation I engine that uses a normally-open switch to signal the ECM when the EGR valve is receiving a vacuum.

grounds the computer to create the pulse, or the sensor, which means that the sensor circuitry is turning the voltage on and off to create the signal.

Examples of a DC frequency pulse generator include the Delco-style MAF and the Hitachi-style MAF.

AC Rotational Pulse Generator

For those familiar with electronic ignition systems, the AC rotational pulse generator is an old friend. Commonly referred to as a pickup coil, the more precise term is variable reluctance transducer. This sensor consists of a permanent magnet, a coil of wire, and a rotating cog. As each of the teeth of the rotating cog enters the magnetic field of the permanent magnet, it causes the field to distort toward it across the coil of wire, resulting in the induction of a voltage. As the tooth of the cog swings past the coil-magnet assembly, the magnetic field is distorted in the other direction, causing a voltage to be induced in the opposite direction. The result of all of this is an AC voltage being generated as each of the cog teeth passes through the magnetic field.

The computer, being a digital device, has a little trouble dealing with this AC signal. Before this signal can be put to use, the computer must convert it to a DC pulse.

This type of sensor is used to sense rotational speed. Included in this category are the distributor pickup coil, vehicle speed sensor

(some), and the DIS (Direct Ignition System) crank sensor.

DC Rotational Pulse Generator

Two types of DC rotational pulse generators are used by Chevrolet. Like the AC versions, the DC rotational pulse generators are used to monitor the rotational speed of a

device. The optical pulse generator is used as a vehicle speed sensor, or VSS, on applications that have a speedometer cable. It uses an LED which shines on rotating shiny metal; the light reflects back to a photodiode. There is a notch in the shiny metal so that, as it rotates, it will cause the sensor to create a pulse which is directly pro-

Variable Resistance: Thermistor

5 Volts

Fuel Injection Computer

A

B

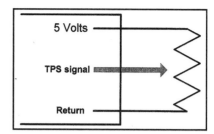

The position of the throttle, and on some applications the EGR valve, is monitored with a potentiometer. These sensing devices are fed a 5-volt reference and a ground. A third wire sends back to the ECM a voltage that varies depending on the position of the device being monitored. Although the MAP sensor is not a potentiometer, it behaves exactly as though it were.

Circuits like this one using a thermistor (temperature-sensitive resistor) are used to monitor both air and coolant temperature. If a voltmeter was connected at either point A or point B, the voltage would be high when the temperature being measured is low. As the temperature being measured increases, the voltage would drop. The ECM monitors the voltage on this circuit and assumes that the lower the voltage, the higher the temperature.

portional to the rotational speed of what is being monitored. This pulse is then sent to the computer.

The second type of DC rotational generator is called a Hall Effects switch. It consists of a permanent magnet sitting opposite a transistor which is extra sensitive to magnetic fields. A set of ferrous metal blades rotates through the gap between the magnet and the transistor, causing the magnetic field to be alternately interrupted between blades. The result is a pulse directly proportional to the speed of rotation.

Types of Hall Effects generators include the C3I crankshaft position sensor and the distributor pickup in some distributor ignition systems.

Both the optical and the Hall Effects pulse generators use three wires. Two wires supply either 5 volts or battery voltage and a ground. The third wire carries the pulse to the computer.

Voltage Generator

Since the oxygen sensor is really the only sensor that creates its own voltage, refer to the section on oxy-

gen sensors in chapter 4 for more details on voltage generators.

The last two types of circuits are actuator circuits.

Normally Grounded

The negative side of the actuator is always connected to ground, and the computer switches voltage to the device on and off to control it. Chevrolet seldom uses a normally-grounded type of circuit.

With a normally-grounded circuit it is possible for a voltage to be induced into the feed wire or the actuator from the computer. This could cause the actuator to energize when it should not. Second, it is impossible for the computer to monitor for damage to the circuit when it is supplying the voltage. All it knows for sure is that it sent voltage out; it has no idea if the voltage was received. Finally, if the computer sends voltage out along a wire and that wire becomes grounded, the computer driver circuit could be overloaded and damaged.

Normally Powered

Chevrolet's method of energizing its actuators is by a normally-powered circuit. The actuator is connected to either battery power or switched ignition voltage all the time; the computer supplies the ground. Since the computer is not supplying power, a grounded wire might blow a fuse or melt a wire—but it will not damage the computer.

Second, the computer is able to monitor the presence of 12 volts to the actuator since it is supplying the ground. Should anything happen to the actuator's power supply, the computer would know it immediately. Finally, there is virtually no way that the actuator could be improperly activated, since it is impossible to induce a ground.

Become familiar with these 11 circuits and you will be familiar with automotive electronics.

The Port Fuel Injection MAF (Mass Air Flow) sensor produces a variable frequency signal. In troubleshooting such a device it must be determined that there is available power, a good ground, and a pulse being sent to the ECM. This pulse can be detected with a dwell meter, a digital tachometer, or a logic probe.

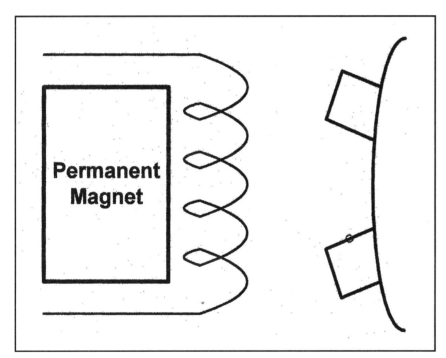

Permanent Magnet

One of the oldest electronic sensing devices is the reluctance pickup coil. The use of this device as an rpm sensor dates back to its use on Chrysler electronic ignition systems of the late 1960s and early 1970s. Today this pickup is still used as the crank sensor on the Chevrolet DIS system and as distributor pickup. Although not related to fuel injection, antilock braking systems use the same device to detect the speed of the wheels. This device produces an AC sine wave which increases in frequency and amplitude as the rotational speed of the reluctor increases. The output signal from this sensor can be measured with an AC voltmeter.

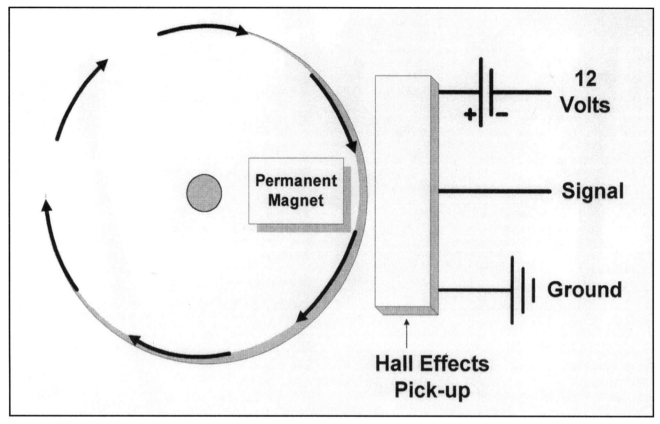

Chevrolet uses two devices that produce a DC pulse to indicate rotational speed. The Hall Effects sensor uses a magnetic field which is interrupted by an armature. The Hall Effects unit is used to measure both rotational speed and position for ignition timing in both distributor and distributorless ignition systems. The optical sensor is used to measure vehicle speed. The output of either one can be measured with a digital tachometer.

The first of two actuator control circuits has the actuating device grounded all the time, with the voltage supply being switched on and off by the ECM to energize and de-energize the device. This is seldom used by Chevrolet since its functioning is difficult for the ECM to monitor.

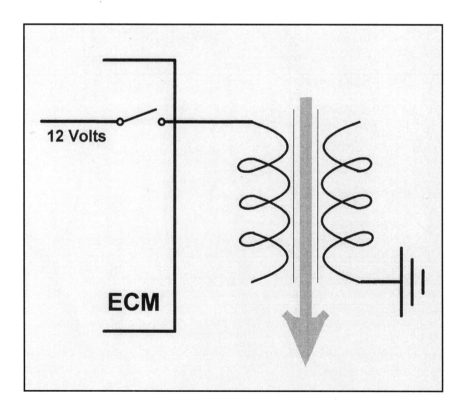

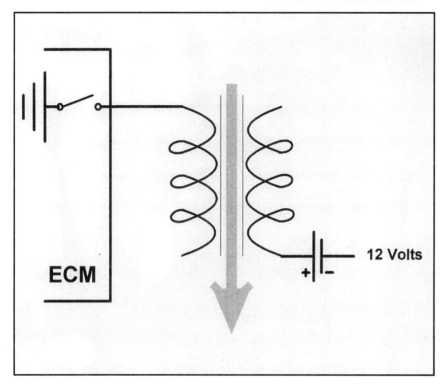

The more common method of controlling an actuator is to send a fused 12 volts to the device all the time and supply a ground through the ECM. This allows the ECM not only to monitor the 12-volt supply to the actuator, but also to confirm that the transistor controlling the device is actually doing its job.

This is the oxygen sensor location on a 1983 2.5-liter Citation. The spark plug wire in the foreground is number four. Of all the sensors used on Chevrolet electronic fuel injection, only the oxygen sensor produces its own voltage.

CONTENTS

DIGITAL
VOLTMETER   28

ANALOG
VOLTMETER   28

AC VOLTMETER   29

TEST LIGHT   29

OHMMETER   29

TACHOMETER   29

DWELL METER   29

DIGITAL LOGIC
PROBE   30

HAND-HELD
VACUUM PUMP   30

FUEL-PRESSURE
GAUGE   30

VACUUM GAUGE   30

LOW-VOLTAGE
OSCILLOSCOPE   30

ROAD SHOCK
SIMULATOR   30

SCANNER   31

TOOLS AND EQUIPMENT

In this chapter we will limit our look at tools to those tools that may be new to the enthusiast, the Saturday afternoon mechanic, or the professional mechanic. We will also look at new ways of using familiar tools. I have made an effort throughout this book to emphasize the simplest, most available tools to perform each diagnostic task.

Digital Voltmeter

The most important tool in working with and troubleshooting modern Chevrolet fuel injection systems can be one of the cheapest and easiest to acquire. Digital voltmeters that are adequate to do the job range in price from a low of about $25 to a high of well over $500. There is an advantage in some of the more expensive meters in that they combine into one tool many of the functions that we will later describe using tachometers and dwell meters.

The digital voltmeter boasts a high-impedance input of 10 million ohms or more. This allows the voltmeter to be connected to very small current flow circuits without affecting the voltage reading. Voltmeters with a low-impedance input tend to rob power from the circuit being tested, causing the voltage readings to be lower than they really are. For this reason the digital meter should be used anytime precise voltage readings are required.

There is a down side to the use of a digital voltmeter, however. Since it is digital, it merely samples voltage and displays it. There are major gaps between these samples. Transient fluctuations are completely missed; there may be a device, such as the throttle position sensor (TPS) potentiometer, which is supposed to create a steadily increasing voltage as the throttle is opened. As the TPS wears, there may be places where the wiper no longer makes contact with the carbon film strip, resulting in a sudden drop in voltage. If the digital volt-meter's sampling did not happen to coordinate with the drop in voltage, the fluctuation, which could be the cause of a major driveability problem, would be missed. For this reason there is a better tool for measuring variations in voltage.

Analog Voltmeter

Where the digital voltmeter displays its reading as digits, the analog voltmeter uses a needle moving across a scale to display its readings. The benefits of the analog voltmeter have suffered much abuse since the introduction of the electronic engine control systems at the end of the 1970s. This is because the majority of inexpensive analog meters have a relatively low-impedance input. As previously mentioned, a low-impedance meter can distort readings. Rumors about technicians ruining ECMs and other components by using analog meters to take measurements are largely exaggerated.

The analog voltmeter will detect fluctuations in voltage much better than the digital voltmeter. When a transient voltage change occurs, it will show up in the analog meter as a fluctuation in the needle.

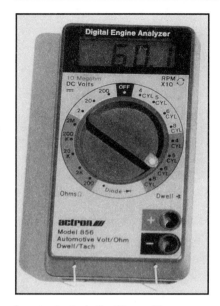

Inexpensive tools like this combine a variety of important functions such as high-impedance voltmeter, tachometer, dwell meter, and ohmmeter all in one handy tool.

Use the analog meter when you are looking for fluctuations in voltage, and the digital when you are looking for precise readings.

Note: Because of the extremely low current output of the oxygen sensor, most analog voltmeters will ground out the oxygen reading. The meter will display 0 volts continuously. Always use a digital voltmeter or a 10 megaohm input impedance analog when taking oxygen readings.

AC Voltmeter

The AC voltmeter has a place in the troubleshooting of electronic ignition and fuel injection systems just as it does in testing a home appliance. The single electronic device that it tests effectively is the reluctance-type pickup. When the reluctor is being rotated at approximately one revolution per second, the AC voltage reading should be between 0.2 and 1.5 volts, depending on the number of teeth.

Test Light

It may seem odd to include such a low-tech testing device in a book about sophisticated fuel injection

systems, but the fact is that it still has a place. Many of the diagnostic flow charts in the Chevrolet fuel injection service manuals suggest the use of test lights to diagnose wiring harness grounds.

Ohmmeter

Ohmmeters are used to measure the resistance in electrical and electronic devices. I am often asked if it is okay to use a low-impedance ohmmeter on a given circuit. Actually, the concept of high impedance and low impedance does not apply to ohmmeters at all. An ohmmeter should always be connected to a component of a given circuit with the power removed from that component. Connecting an ohmmeter to a powered circuit can damage both the tester and the circuit.

Special warning: Since the ohmmeter contains its own power source that it uses as a reference to determine the resistance of what is being measured, and since the oxygen sensor is a voltage generator, connecting an ohmmeter to an oxygen sensor will destroy the O_2 sensor.

Tachometer

Still handy for checking engine speeds, a digital tachometer also has other uses. One of the new diagnostic instruments needed on fuel-injected cars is a frequency counter. Since even most professional fuel injection technicians do not own a frequency counter, let me suggest the use of a digital tachometer.

The tachometer measures the number of primary ignition pulses per minute and mathematically converts the number of pulses into the number of crankshaft rotations. It sends that calculation to the display as crankshaft rpm. By doing some very simple math, we can get a frequency reading in cycles per second, or hertz.

You'll need to switch the tachometer to the four-cylinder scale, whether the engine you are working on is a four-cylinder, six-cylinder, or eight-cylinder (remember, we are counting pulses per second, not

Many sources advocate disposing of your analog voltmeter for the purposes of diagnosing electronic fuel injection systems. However, an analog voltmeter is superior to a digital in its ability to read smooth and progressive changes in voltage.

crankshaft rpm). This technique will work equally well on the six- or eight-cylinder scale, but the math is more complex. Connect your tachometer to the wire (usually the ground side of an actuator circuit or the output of an appropriate sensor) you want to test and a good ground.

Observe the reading on the tachometer and divide by 30. That will be the reading in hertz.

$$\frac{4{,}800 \text{ rpm}}{30} = 160 \text{ Hz}$$

An analog tachometer is not recommended for this purpose, since it requires a degree of precision not found in most analog units.

Dwell Meter

Since the advent of electronic ignition and HEI (High Energy Ignition), the dwell meter has been gathering dust. No longer do we make adjustments on points during a tune-up. However, as the tachometer was actually measuring the frequency of the primary ignition system, the dwell meter was measuring its duty cycle; therefore, when duty cycle needs to be measured, the dwell meter again becomes a handy tool.

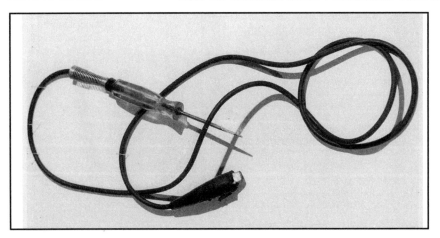

The basic test light can be used to troubleshoot actuator power supplies, such as those for the injectors. It is important to remember that none of the sensor circuits will provide enough potential current flow to make a test light glow. Therefore, using a test light without keeping this in mind can result in an incorrect diagnosis.

The ohmmeter can be used to test the resistance in both sensors and actuators. A typical place it might be used is to measure the resistance in this 1983-vintage coolant temperature sensor. Note: Never connect an ohmmeter to a coolant temperature sensor.

Remember that the duty cycle is a measurement of the relationship of the on-time and the off-time. Place the dwell meter on the four-cylinder scale; again, do not be concerned with how many cylinders the engine has that you are working on. Connect your dwell meter to the wire on which you want to read the duty cycle, and observe the reading. Multiply the reading by 1.1.

Note: In reality, there will almost never be a time in automotive troubleshooting where that degree of precision is required. Simply using the observed reading will suffice.

Digital Logic Probe

A digital logic probe is like a high-tech, high-impedance test light. Designed to register whether a voltage is above or below a certain point, this logic probe can also be used to detect the presence of a rhythmic change in voltage, or a pulse.

Hand-Held Vacuum Pump

A hand-held vacuum pump can be an invaluable addition to your tool box. Not only will it come in handy in testing MAP sensors, EGRs, and vacuum-controlled actuators, but it also handy for testing such low-tech items as vacuum advance units.

Fuel-Pressure Gauge

If you do not have a fuel-pressure gauge, you might as well forget about troubleshooting fuel injection systems. All good troubleshooting begins with a fuel-pressure test. The fuel-pressure gauge that is combined with your vacuum gauge will no longer fit the bill, however; you will need a gauge capable of accurate readings of up to 75 psi (pounds per square inch). A variety of fittings and adapters will also be necessary. Shop around a little. The prices on these gauges with fittings can run anywhere from $100 to $1,000. A little ingenuity and a trip to a local hydraulics or air-conditioning supply store could probably yield a more than adequate gauge and a significant savings.

Vacuum Gauge

An ancient workhorse, the vacuum gauge is still as valuable as it ever was for detecting mechanical problems with the engine and troubleshooting mis-routed and damaged vacuum circuits.

Low-Voltage Oscilloscope

The low-voltage oscilloscope, also a valuable tool, is new to the arsenal of automotive diagnostic weaponry. This engine analyzer scope has been used for decades to troubleshoot primary and secondary ignition systems. Early in the 1980s the manufacturers of these analyzers saw the need to test patterns and wave forms at much lower voltages than what you would experience in the ignition system. They began to incorporate low-voltage functions in their professional-quality engine analyzer scopes.

You can purchase an oscilloscope of this type, however, for considerably less than the thousands of dollars a pro shop has to invest. Through an electronics hobby store they cost as little as $500. Used oscilloscopes can cost as little as $100.

Keep in mind, though, that this type of scope cannot be used to analyze either primary or secondary ignition.

Road Shock Simulator

One of the more fun, though severely abused, test tools is the road shock simulator. This tool consists of a piece of wood approximately 18 inches long and 3/4 inch in diameter.

$$Hz = \frac{RPM\ (4cyl)}{30}$$

This formula reverses the mathematical conversion performed by the tachometer in changing frequency into rpm. The Delco MAF produces a variable frequency square wave signal that can be measured with a frequency counter or with a digital tachometer and the formula. At an idle, the rpm reading from the MAF should be in the neighborhood of 1,000. Using this formula, frequency would be about 33 Hz.

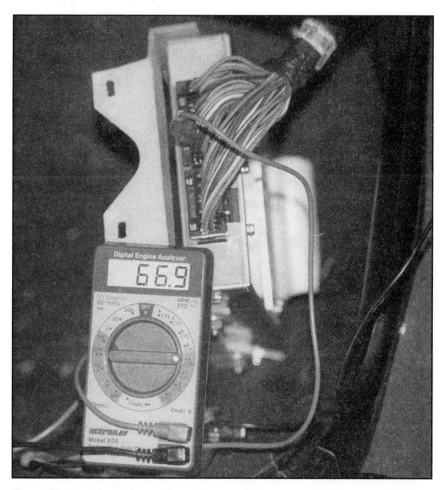

The usefulness of the dwell meter began to wane with the advent of electronic fuel injection in the mid-1970s. A digital dwell meter can be used to test for the presence of a pulse, or to measure a variable duty cycle. Here, the dwell meter is connected to the EST wire monitoring the variable duty cycle of the timing control signal.

The weight of this tool falling about 1 inch (with no muscular assistance) can help to spot intermittent open circuits in various sensors and actuators.

If the simulator in the drawing looks amazingly like the handle from a plumber's friend, that's because that is precisely what it is.

Scanner

Probably one of the most essential tools that is going to be involved in troubleshooting or fine-tuning on a regular basis is the diagnostic scanner. A scanner connects to the ECM's diagnostic connector, known as an ALDL connector, and translates computer code from the ECM into digital information about what the computer is seeing, thinking, and doing.

Scanners come in all sizes, price ranges and levels of user friendliness. Scanners are incorporated into engine analyzers costing tens of thousands of dollars, and in hand-held units such as those marketed by Snap-on, OTC of Owatonna, Minnesota, and others at between $1,000 and $2,000. A company named Rinda Technologies markets a PC-based computer program called DIACOM that will work on most any MS-DOS home computer with a serial port. The program sells with the necessary serial interface cable for about $300. (See appendices for supplier addresses.)

Although a scanner is not a necessary tool for the average enthusiast, it is definitely a valuable tool. Its real value comes from its use in conjunction with flow charts for diagnosing problems related to trouble codes.

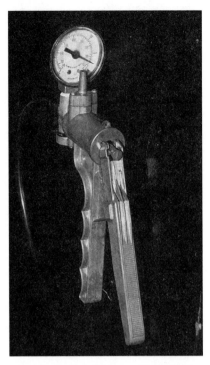

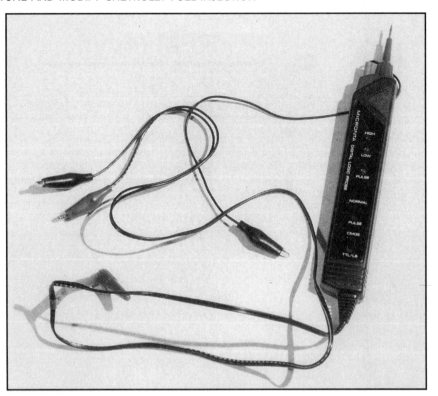

Priced from under $20 to well over $100, the hand-held vacuum pump can be an invaluable addition to your tool box. With the key on and engine off, connect a hand-held vacuum pump to the MAP sensor. With no vacuum applied the MAP sensor output voltage should be about 4.5 volts. As you slowly apply a vacuum, the voltage should change a proportional amount. The vacuum pump can also be used to test the EGR valve diaphragm and vacuum-controlled actuators.

Although usually reserved for the electronics technician, the digital logic probe can detect the presence of pulses and be used as a high-impedance test light to check for the presence of a reference voltage in a sensor circuit.

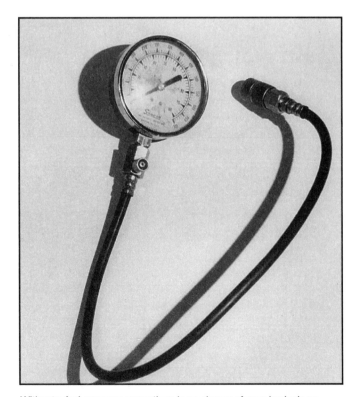

Without a fuel-pressure gauge there is no chance of even beginning a competent diagnosis. If the fuel pressure is not correct, the rest of the injection system cannot operate properly. Since the ECM does not monitor fuel pressure, it has to be able to assume that the fuel pressure is correct.

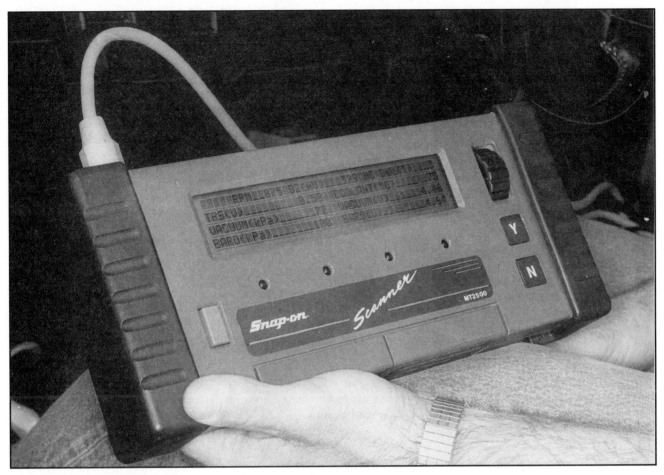

One of the most sophisticated diagnostic tools for Chevrolet fuel injection is the diagnostic scanner. This device connects to the ALCL connector and interprets data being sent to that terminal by the ECM. This allows the technician to look at the inputs and some of the thinking processes of the ECM.

```
                              DIACOM

VEHICLE TYPE: 1988 CHEVROLET        ENGINE TYPE: 2.8 Liter V6

ECM MODE STATUS: ALDL               ECM PROM ID: 991

Engine Speed................  850 RPM   Spark Advance...............  22.2 DEG
Desired Idle Speed..........  850 RPM   Knock Retard................   0.0 DEG
Vehicle Speed...............    0 MPH   Spark Control Counts........     2 #
Coolant Temperature.........  176 F     Integrator..................   126 #
Start-up Coolant Temp.......   75 F     Block Learn Multiplier......   118 #
Manifold Air Temperature....   52 F     Block Learn Cell Number.....     0 #
Throttle Sensor Voltage.....  0.63 VDC  Battery Voltage.............  14.2 VDC
Throttle Position...........    0 %     Fuel Pump Relay Voltage.....  14.2 VDC
MAP Sensor Voltage..........  3.94 VDC  Engine Running Time.........    10 Min
Barometric Pressure.........   26 Hg    Catalytic Conv. Temp........   691 F
Oxygen Sensor Voltage.......  597 mV    EGR Duty Cycle..............     0 %
Oxygen Sensor Transitions...  110 #     EGR Pos. Sensor Voltage.....  0.86 VDC
Idle Air Position...........    0 #     EGR Diagnostic Switch.......   On
Air Flow Rate...............    4 GPS   Fuel Mixture Status.........   Rich
Injector Pulse Width........  1.6 mS    Closed Loop Status..........   On

Configure  Disk  Exit  Info  Link  Mode  Options  Print  Trbl_codes
Establish communication with vehicle
COMMAND=> USE ↔ ARROW KEYS or SPACE BAR TO SELECT OPTION THEN PRESS "Enter"
```

The diagnostic scan tool allows the technician to peer into the thinking processes of the ECM. Displayed on the scanner screen are the input voltage of the sensors and the engine operating conditions. The scanner illustrated here is a PC-based computer program that displays up to 30 data fields simultaneously. Other hand-held scanners will display 2 to 9 fields of data.

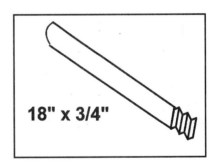

18" x 3/4"

A little bit of fun and a little bit of seriousness: The road shock simulator. Consisting of nothing more than the handle of a plumber's friend (plunger), the road shock simulator is used to detect intermittent faults in a sensor. By allowing the unassisted weight of the stick to fall on the sensor, it provides a shock similar to what the sensor might experience when the car is driven over less than perfect roads.

33

CONTENTS

FUEL SYSTEM	**35**
ELECTRONICS	**39**
ELECTRONIC CONTROL MODULE	**45**
SPARK CONTROL	**50**
EXHAUST GAS RECIRCULATION CONTROL	**51**
MAF BURN-OFF	**53**
AIR INJECTION SYSTEM	**53**
TRANSMISSION CONVERTER CLUTCH	**53**
AIR CONDITIONER COMPRESSOR CLUTCH	**53**
COOLING FAN	**53**

CHAPTER FOUR

HOW CHEVROLET FUEL INJECTION WORKS

As discussed in chapter 1, Electronic Fuel Injection is not a new concept, nor is it a mysterious one. A modern fuel injection system has the same objectives as does a carbureted engine. There are six essential tasks of a fuel-injected engine: start-up enrichment; cold-running enrichment; idle speed and air-fuel ratio control; cruise air-fuel ratio control; load enrichment; and acceleration enrichment.

A carburetor performs each of these tasks without the assistance of additional sensors or computers. Even though in the late 1970s and on into the 1980s computers were added to carbureted cars, the carburetor was able to function even when the computer was removed from the car. The computer on a carbureted car was there to fine-tune the air-fuel ratio, not to make the carburetor function. The carburetor has within itself the ability to meet the essential tasks of a fuel supply system.

Carburetor Functions

Engine task	Carburetor function
Start-up enrichment	Choke full on
Cold-running enrichment	Choke partly on
Idle speed and air-fuel ratio control	Idle circuit
Cruise air-fuel ratio control	Cruise circuit
Load enrichment	Power valve
Acceleration enrichment	Accelerator pump

Fuel Injection Functions

Engine task	Component/sensor function
Start-up enrichment	Cold-start valve or coolant sensor
Cold-running enrichment	Coolant sensor
Idle speed and air-fuel ratio control	Idle air control, MAP or MAF
Cruise air-fuel ratio control	MAP or MAF, oxygen sensor
Load enrichment	TPS, MAP or MAF
Acceleration enrichment	TPS, fuel-pressure regulator

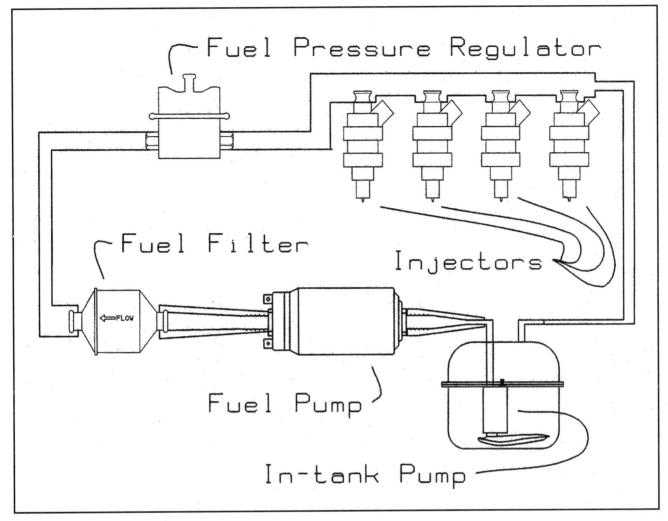

The driver, or output device, circuits have less resistance in them and therefore a greater current flow. These larger current flows are necessary to generate magnetic fields large enough to activate the solenoids and relays that carry out the operational functions of the engine. Corrosion and other sources of resistance can affect the strength of the magnetic fields and therefore the functioning of the output circuits.

This chapter will look at the operating principles of Chevrolet fuel injection systems. We will look at the design, construction, and function of each of the components, and how they work together to deliver performance, driveability, and low emission levels. An understanding of how the ECM acquires, interprets, and utilizes information that it takes in will better help the technician or enthusiast successfully troubleshoot or modify GM fuel injection systems.

We will cover several major subsystems in the GM fuel injection system. These subsystems include: Fuel; Electronics; Electronic Control Module; Air Induction System; Spark Control; Exhaust Gas Recirculation Control; MAF Burn-Off; Air Injection System; Transmission Converter Clutch; Air Conditioner Compressor Clutch; and Cooling Fan.

Fuel System

In this section we will look at the fuel components of Throttle Body Injection (TBI), Port Fuel Injection (PFI) and Tuned Port Injection (TPI) systems. Where necessary, distinctions will be made concerning these components, but I will try to simplify by treating similar components as the same. Also, where appropriate, a from-the-trenches diagnostic tip will be offered.

Fuel Tank

The fuel tank on an electronic fuel-injected car is not significantly different from its carbureted counterpart. The outbound line from the tank is larger to accommodate the increased volume of fuel required by the injection system. There will also be a large return line. At times, 90 percent or more of the fuel passing through the fuel pump will travel completely through the fuel system and return to the tank.

Fuel Pump

Chevrolet uses two types of fuel pumps on its fuel-injected cars. One is a low-pressure pump which is used

A TBI fuel pump from an early 2.5-liter Citation. This is an electric vane-type pump. Located inside the fuel tank, the pump is capable of moving far larger quantities of fuel through the injection system than the engine would ever use. On PFI applications, the pump is also mounted inside the tank but is larger and produces a higher volume.

The fuel filter is probably the only maintenance item that cannot be changed too often. Offering the only real defense against contamination in the injectors, a restricted fuel filter can cause a wide range of symptoms from low power to stalling at an idle. The best test procedure for the condition of the fuel filter is simply to replace it.

on the TBI and CFI (Crossfire Injection) applications. The other is a high-pressure pump that is used on the PFI and TPI applications. It is actually inaccurate to refer to these pumps as high pressure and low pressure because pumps do not really create pressure, they merely supply a volume of fuel. It is the fuel-pressure regulator restricting the volume of fuel returning to the tank that creates the pressure.

Both the low-pressure TBI and high-pressure PFI pumps are located in the fuel tank. The TBI system uses a high-volume DC motor vane-type pump that rotates at about 3,500 rpm. The PFI systems use a roller vane pump. Both of these pumps are extremely efficient at pushing fuel through the injection system, but are poor at pulling the fuel from the tank. For this reason their location inside the tank is ideal.

The design of the pump is such that fuel pulled in through the inlet passes through the electric motor portion of the pump. This fuel passing through the motor acts as a coolant and lubricant for the pump. There is a "sock" filter on the inlet side of the pump that prevents large, hard particle contamination from getting into the pump. The pump can also be damaged by foreign material that can easily pass through this filter. Such things as water and alcohol can do severe damage. If you're careful where you buy your

fuel, you and your fuel pump will have a long and happy life together.

The fuel pump is equipped with a check valve on the outlet side. This valve prevents fuel in the system from draining back into the tank. If the check valve should become defective, it will show an extended-start symptom. After the car sits for awhile, the engine will have to be cranked for several seconds before it will start. The cure for this problem is replacing the pump. Keep in mind, however, that a loss of residual fuel volume is not the only possible cause of an extended-start symptom.

Fuel Filter

The fuel filter is the main line of defense against hard particle contamination in the injectors. These filters consist of a fine, paper-mesh filter in a metal can capable of filtering out particles as small as 10 microns (0.0004 inch).

Over the years I have seen many inappropriate installations of parts. A fuel injection fuel filter is one of the most potentially dangerous of these misapplications. I remember a

Volkswagen van where the owner had installed a universal fuel filter normally used on VW carbureted engines. This filter was designed to operate in a system where the pressure is only about 3 or 4 psi. The van was fuel injected with a system pressure of over 30 psi. The filter had ballooned to about 25 percent larger than its original size, with white stress lines running its entire length. This was a potential rolling coffin of fire. Do not attempt to adapt an incorrect filter to your car. Maximum pressures, flow rates, and differences in fittings make it either impractical or dangerous.

Some filter manufacturers stamp an arrow on the side of the filter to identify the direction of flow; others do not. When you install a new filter, make sure that it is installed in the proper direction.

The fuel filter is one maintenance part that cannot be changed too often. It is much cheaper than replacing contaminated injectors and pressure regulators. Replace the fuel filter at every major tune-up, regardless of the suggested service interval

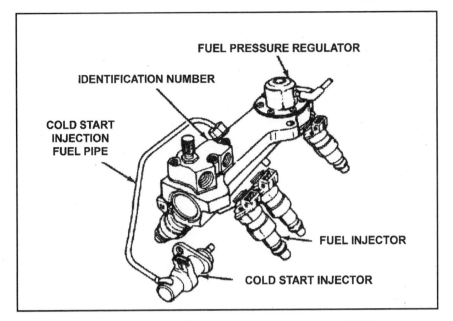

This is the fuel rail from an early 2.8-liter engine as might be found in a 1986 Celebrity. Notice that the fuel rail itself is a cast unit with a Schraeder valve provided for easy access when testing fuel pressure. Unique to this engine as well as the 5.0- and 5.7-liter V-8 is the cold-start injector. This extra injector sprays fuel while the engine is being cranked when cold to enrich the mixture. Once the engine starts and the ignition switch is released, the cold-start injector shuts off.

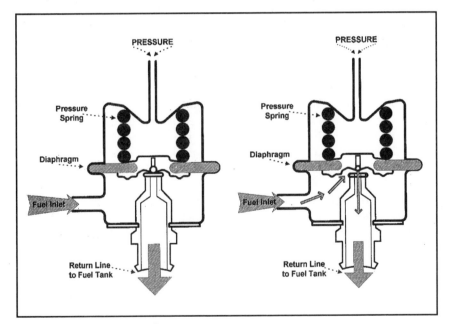

The fuel-pressure regulator consists of a spring-loaded diaphragm that is deflected to open a valve as the fuel pressure increases. Manifold pressure is used on PFI and TPI applications to increase the fuel pressure during acceleration.

in your owners manual. Also, replace the fuel filter anytime other components in the fuel system such as a pump, injectors or pressure regulator are replaced.

Fuel Rail

A line from the fuel filter runs to the engine and attaches to the fuel rail. Chevrolet's fuel rail is usually cast aluminum. All of the fuel-system components on the engine are attached to it. Although it is less evident, TBI systems also have a fuel rail—the chambers within the throttle body assembly that connect the inbound fuel line, the pressure regulator and the injector(s).

Fuel-Pressure Regulator

As mentioned, pumps do not create pressure but rather create volume; it is the fuel-pressure regulator that is the primary restriction in the system to create pressure. Proper fuel pressure is critical to maintaining the correct air-fuel ratio. If the fuel pressure is incorrect, then the electronic controls and sensors will find it difficult or impossible to meter the correct amount of fuel to provide the proper mixture.

Most PFI and TPI systems run about 35 psi at an idle. There is a vacuum line connected to the top of the pressure regulator. The vacuum line is connected to the manifold vacuum, and when the throttle is opened by the driver, the manifold vacuum drops, which causes the fuel pressure to increase.

This concept might be a little easier to appreciate when you realize that, scientifically speaking, there is no such thing as a manifold vacuum. What actually exists in the intake manifold with the engine running is pressure. This pressure is known as manifold pressure and is lower than atmospheric pressure. Since when we measure it we are standing in atmospheric pressure, it feels like a vacuum. This pressure increases as the volume of air entering the manifold increases. Therefore, when the throttle valve is opened, manifold pressure increases, which means the pressure traveling through the vacuum hose to the fuel-pressure regulator increases as well. As the pressure in this line increases, it in turn causes the fuel pressure to increase.

The increase in fuel pressure as the throttle is opened averages 5 to 10 psi and is intended to act as an accelerator pump enriching the mixture during acceleration. Increasing the fuel pressure is also necessary to maintain the proper pressure differential across the tip of the injector as the intake-manifold pressure increases. It should be noted that anytime a given throttle setting is maintained, manifold pressure will decrease a bit; therefore, the fuel pressure will begin to drop toward the idle pressure.

Throttle Body Injection applications control fuel pressure between 9 and 12 psi. Unlike Port Fuel Injection applications, the TBI fuel-pressure regulator is rebuildable. Rebuild kits are available both through the dealer and aftermarket suppliers. Caution should be used when disassembling the fuel-pressure regulator, as it contains a long spring under sufficient pressure to be a safety hazard if released carelessly.

Fuel pressure does not change as the throttle is opened on TBI and CFI. Additional pressure is not needed since the fuel is being sprayed ahead of the throttle plates into the relatively constant pressure of the atmosphere.

Cold-Start Injector

The cold-start injector, also known as the cold-start valve, is used only on some 2.8 liter PFI systems, and the 5.0 liter and 5.7 liter TPI. This injector opens only when the engine is being cranked and the temperature of the engine is below 100 degrees Fahrenheit. As it opens, it provides additional fuel to the intake manifold to enrich the mixture for cold starting. Even at the coldest temperatures it will spray for only about 12 seconds. The closer the temperature is to 100 degrees, the

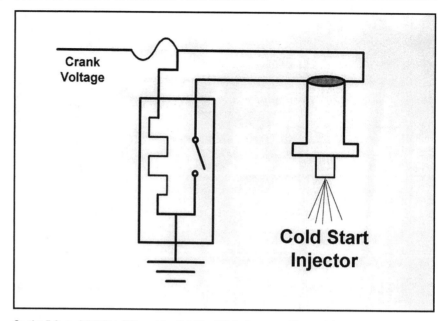

On the 5.0- and 5.7-liter TPI and the 2.8-liter PFI that use a cold-start injector, power is supplied to the injector by the start position of the ignition switch. The cold-start injector is then grounded through the thermo-time switch, which consists of a bimetal switch and an electric heater. If the coolant temperature is less than 95 degrees Fahrenheit, the bimetal contact will be closed until the electric heater heats the switch over a period of a few seconds.

shorter time it will spray. Control of the cold-start valve is by means of the thermo-time switch.

Thermo-Time Switch

The Thermo-time switch is found only on the applications that use a cold-start valve. This is a normally-closed temperature-sensitive bimetal switch that typically supplies a ground for the cold-start valve through one of its two electrical connections. When the switch is heated to about 95 degrees Fahrenheit the bimetal flexes, causing the cold-start to lose its ground. There are two methods of heating the switch.

The thermo-time switch is screwed into the water jacket so that engine temperature will open the switch. Another way involves the second electrical connection. This connection feeds current to an electrical heater which will heat the bimetal to the point where the switch opens in 5 to 12 seconds.

Injectors

The modern electronic fuel injector is a normally-closed, solenoid-operated valve. On Chevrolet applications the injector is connected to a 12-volt power source, and the ECM connects it to ground in order to energize and open the injector. There are two major categories of injectors used by Chevrolet. The first is the top-feed or high-pressure injector. This injector is used exclusively on the PFI and TPI applications. They are connected to the fuel rail and sealed by an O-ring. The other end of the injector rests in the intake manifold where vacuum leaks are also sealed by an O-ring. Fuel pressure on these applications average 35 psi at an idle.

The other type of injector is known as a side-or bottom-feed injector. These are used exclusively on the TBI applications at a typical fuel pressure of between 9 and 12 psi. Only one of these injectors is used to fuel an entire four-cylinder engine, and two are used on the TBI-equipped V-6 and V-8 engines.

The injectors on the TBI applications, like those of the TPI and PFI applications, are connected to switched ignition 12 volts and are

The coolant temperature sensor (CTS) monitors the engine temperature. As the temperature of the engine increases, the resistance of the CTS decreases. Checking its resistance with the engine thoroughly warmed up should yield a reading of about 170 to 250 ohms (ideal is 185 ohms at 212 degrees Fahrenheit).

This injector is typical of PFI injectors. The ECM enables a current flow through the injectors lasting for about 1 millisecond. As current flows through the solenoid, a valve is opened which allows fuel to flow through the injector. A typical length of time for the injector to be open without the engine under load at an idle is about 1 to 1.5 milliseconds.

Manifold Absolute Pressure (MAP) is an indicator of engine load as well as the velocity of air entering the intake manifold. Beginning with a signal voltage of about 1.5 at an idle and increasing to around 4.0+ voltage under full load or at wide-open throttle, the MAP is used on all TBI and some PFI applications.

grounded by the ECM to open them.

The ECM controls the energizing of the injectors with transistors or quad drivers. On the PFI and TPI applications, the injectors are synchronized to primary ignition and open simultaneously once every crankshaft revolution. At first this may sound a little inefficient or wasteful, but remember that the fuel sprayed on top of each of the intake valves is the correct amount to fuel that cylinder and will merely sit on top of the valve until the valve opens, allowing the air rushing into the cylinder to carry the fuel.

On TBI and CFI, an injector opens every time there is a primary ignition pulse. This means that on a one-injector TBI system, the injector opens every time a spark plug fires. On the two-injector applications, the injectors will alternate so that each injector opens every other time a spark plug fires.

Fuel Lines

All of these components are connected by hoses or steel lines. The hose that runs from the fuel pump to the fuel rail is called the supply line. Somewhere in the supply line is the fuel filter. The line that runs from the fuel-pressure regulator back to the fuel tank is called the return line. Although in most cases these lines or hoses are ignored during the troubleshooting procedure, it should be kept in mind that they can develop problems such as restrictions, kinks, and collapsed portions. Because of the higher pressures and volumes associated with fuel injection systems when compared to carburetors, do not use standard fuel line when replacing hoses on a fuel-injected car. Always use high-pressure-rated fuel injection line.

Electronics
Sensor Coolant Temperature

The coolant temperature sensor, or CTS, is a resistor known as a negative temperature coefficient thermistor. This type of resistor responds dramatically to changes in temperature. At 40 degrees Fahrenheit, the resistance is about 100,700 ohms. As the temperature increases, the resistance decreases; as the resistance decreases, the voltage of the coolant temperature circuit as measured by the ECM decreases. At 212 degrees Fahrenheit, the resistance is around 200 ohms. When the engine is cold, the wire carrying the 5-volt reference will have about 3 to 4 volts on it, depending on the exact temperature of the engine coolant. As the engine warms up this voltage will drop, pulled low by the decreasing resistance in the coolant sensor.

The coolant temperature sensor replaces the choke of a carbureted car. When the ECM detects a high voltage on the CTS wire, it increases the pulse width to the injectors, which enriches the mixture. In addition to enriching the air-fuel ratio, it also causes the EMC to permit additional timing advance when the indicated temperature is low.

Here's a track trick. Installing a 640 ohm resistor across the terminals of the coolant sensor wiring harness with it disconnected from the coolant sensor will trick the ECM into believing that the engine is not quite warmed up yet—or at about 145 degrees Fahrenheit. The ECM will respond by enriching the mixture and allowing more timing advance sooner. The net result is an increase in track performance for less than 50 cents.

A word of caution about this trick: Since on some applications the radiator cooling fan is controlled by the ECM's perception of coolant temperature, this trick should be used only for short runs at the drag strip. Prolonged use may cause severe overheating and engine damage since the ECM would never see the coolant sensor indicating a hot enough temperature to turn on the radiator fan.

Manifold Absolute Pressure Sensor

The Manifold Absolute Pressure, or MAP, sensor is a piezo-resistance device that changes a 5-volt reference voltage in response to changes in manifold pressure. As manifold pressure increases, the voltage from the MAP sensor also increases.

Before going any further with this explanation, it should be noted that most technicians and car hobbyists are used to thinking in terms of manifold vacuum. Scientifically, there is never a vacuum in the intake manifold. What is in the manifold is a pressure that is lower than atmospheric pressure. We are also used to thinking that when the throttle is opened the manifold vacuum decreases. Actually, the manifold pressure is increasing and because the pressure is also increasing, the reading on a vacuum gauge drops.

When the throttle is opened or the engine is put under a load in some other fashion, the manifold pressure will increase. As the manifold pressure increases, the voltage output from the MAP sensor also increases. At or near seal level, an idling engine will cause the voltage signal from the MAP sensor at between 1.2 and 1.9 volts, usually averaging about 1.5. As long as the idle voltage is in this neighborhood and the voltage increases as the engine load increases (such as when you snap the throttle), and decreases as the engine load decreases, then the MAP sensor is functioning normally.

All TBI applications use a MAP sensor as the primary means of detecting both engine load and airflow into

Shown here disassembled, the Bosch MAF uses a hot wire to measure how much air is entering the engine. This type of MAF is used exclusively on the Tuned Port Injection systems. Producing a variable voltage, the Bosch MAF signal voltage starts low (less than 1 volt at an idle) and increases to over 5 volts as the mass of air entering the engine increases. (Note: Disassembling the MAF like this destroys it.)

the engine. Some of the PFI engines use a MAP sensor such as the Generation II 2.8-liter, the 1.8-liter turbo, and the 2.0-liter turbo.

On the turbo applications, it should be noted that all of the output voltages for a given pressure are cut in half, making the idle voltage about 0.75 volts. This is because the MAP sensor on a turbo not only has to measure pressures below atmospheric, but it also boosts pressures above atmospheric pressure.

The ECM uses the MAP sensor information for the control of two major systems. First, the MAP sensor signal is used to measure the flow of air into the engine on applications that are not equipped with a Mass Air Flow sensor. Second, it is used to measure engine load in order to retard the ignition timing when the engine comes under a load.

Mass Air Flow Sensor

Most PFI and TPI applications use a Mass Air Flow, or MAF, sensor to measure the flow of air into the engine. Exceptions to this would be the aforementioned 1.8- and 2.0-liter turbocharged applications, as well as the 1989 and later Generation II 2.8-liter engine, which uses a MAP sensor. For the 1988 Generation II 2.8-liter, the MAP sensor is used along with the MAF. The MAF provides the ECM with measurements of the amount of air being drawn in by the engine, while the MAP is used to measure the load on the engine. A service bulletin that came out in late 1989 points out that a PROM (Programmable Read Only Memory) update is available which ignores the MAF sensor. It can only be assumed from this that there have been problems with this MAF sensor.

The Delco MAF is used on most PFI applications. Although it functions in a manner similar to the Bosch style, its output signal is a variable frequency. The output signal starts low, around 32 Hz with the engine shut off, and increases to a maximum of about 150 Hz as the air mass entering the engine increases.

This is the Delco 2 MAF, as installed on a 1994 Corvette LT1 TPI engine. The Delco 2 MAF has proven to be significantly more reliable and more accurate than the original Delco MAF. *Jason Scott*

Currently, three different types of Mass Air Flow sensors are being used by General Motors, Bosch, Delco, and Hitachi.

Bosch MAF Sensor

The first and original Mass Air Flow sensor is the Bosch, used on Tuned Port Injection applications. This mass airflow sensor produces a voltage that varies with the airflow into the engine. With a signal voltage range of 0–5 volts, the lower the voltage, the smaller the perceived airflow.

The Bosch MAF uses a heated wire to detect the mass of incoming air. This wire is made of platinum and is approximately the thickness of a human hair. A small "computer" mounted on the side of the sensor applies a current flow to the wire. This heats the wire to approximately 100 degrees Fahrenheit above the temperature of the incoming air. The temperature of the incoming air is measured by a thermistor located across the bore of the sensor.

As the engine is started and incoming air begins to pass through the sensor, it cools the heated wire. The little computer on the side of the MAF detects this loss of heat as a change in the voltage drop across the wire and increases the current flow to maintain the temperature of the wire. The ECM is then sent a voltage signal that is directly proportional to this current flow.

What makes the MAF potentially a near-perfect air-measuring device is the fact that it can measure four characteristics of the inbound air charge: volume, air temperature, barometric pressure/altitude, and humidity.

First, as the volume of the air entering the engine increases, the tendency will be for the heated wire to be cooled more. As the MAF increases the current flow through the wire to maintain the temperature, the voltage output to the ECM is increased and the ECM adds more fuel through the injectors to match the additional air.

The second characteristic of inbound air charge is air temperature. Obviously colder air will have an increased cooling effect on the wire. Colder air is also denser, requiring an increase in the amount of fuel needed for a given volume of air. Since the colder air cools the wire more, the computer must increase the current flow through the heated wire to maintain the proper temperature. This increase in the current flow is matched by the ECM, and the ECM in turn increases the amount of fuel passing through the injectors to match the needs of the colder, denser air.

Third, as altitude increases, the barometric pressure drops. Barometric pressure can also change as the weather changes. As the barometric pressure drops, the density of the air also drops, requiring less fuel at a given air volume to maintain the proper air-fuel ratio. This less dense air will also have less of a cooling effect on the heated wire as it passes through the MAF. The ECM will respond by decreasing the injector pulse width, thereby decreasing the amount of fuel entering the combustion chamber.

Humidity, the fourth characteristic of the air charge, also has an effect on the density of the air. When the air is humid, the moisture tends to push the air molecules father apart, thereby decreasing air density. This less-dense air requires less fuel per unit of air volume, but also has less of a cooling effect on the heated wire. The air-fuel ratio is automatically maintained.

The Bosch MAF is equipped with a self-cleaning function. Whenever the engine has been operated in the closed loop mode, the heated wire will be heated red-hot for approximately 1 second when the engine is shut off. This burn-off occurs approximately 4 seconds after the engine is shut off. The purpose is to clean off any road oil or dirt that may have stuck to it while the car was being driven.

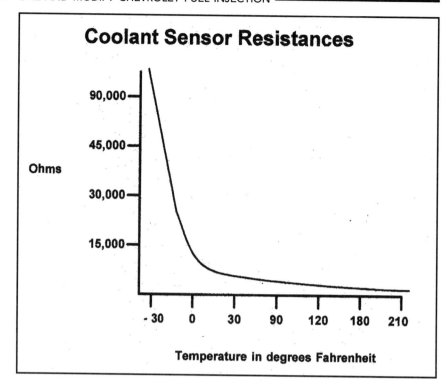

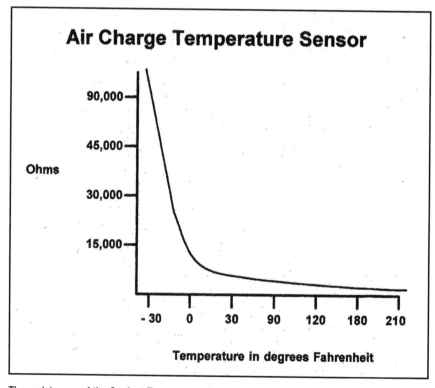

The resistances of the Coolant Temperature Sensor and the Manifold Air Temperature sensor (also known as the Air Charge Temperature sensor) are exactly the same at any given temperature. Beginning at over 100,000 ohms at -40 degrees Fahrenheit, the resistance drops to less than 200 ohms at 212 degrees Fahrenheit.

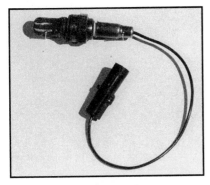

The oxygen sensor monitors the air fuel ratio by measuring the amount of oxygen in the exhaust gases. When the oxygen content increases, the sensor voltage drops to less than 450 millivolts and the ECM assumes that the engine is running lean. When the oxygen content of the exhaust gases decreases, the oxygen sensor voltage increases and the ECM assumes that the engine is running rich. Measure the oxygen sensor voltage with either a high-impedance voltmeter or a scanner. Caution: Never connect an ohmmeter to an oxygen sensor.

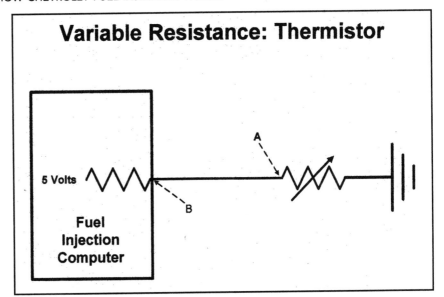

Variable Resistance: Thermistor

Circuits like this one using a thermistor (temperature sensitive resistor) are used to monitor both air and coolant temperature. If a voltmeter was connected at either point A or point B, the voltage would be high when the temperature being sensed is low. As the temperature being measured increases, the voltage would drop. The ECM monitors the voltage on this circuit and assumes that the lower the voltage, the higher the temperature.

Delco MAF Sensor

The Delco MAF works essentially the same way as the Bosch type, but there are some important differences. The Delco MAF doesn't use a heated wire to measure the incoming air; instead, it uses a piece of Mylar film with a copper conductor in it. The computer mounted on the side of the sensor maintains a temperature on the film and converts its control of current through the film into a variable frequency. The frequency varies from 32 to 150 Hz with the lower frequencies indicating a smaller airflow.

There is no burn-off function on the Delco MAF.

Delco 2 MAF Sensor

In the mid-1990s, Chevrolet began to use a new Mass Air Flow sensor built by Delco. Unlike the Delco MAF used by other divisions of GM during the 1980s, this MAF has proven to be very dependable. The unit is a tube located in the main air induction system, similar to the Bosch MAF. All of the air destined for the engine must pass through the MAF. Working similar to other Mass Air Flow sensors, the second-generation MAF puts out a variable frequency that ranges from several hundred Hertz at idle to several thousand at full throttle.

Hitachi MAF Sensor

The Hitachi MAF, used by other GM divisions, is similar to the Delco in that it creates a frequency, but the frequency is much higher. It differs from both the Delco and the Bosch in that the heated sensing element lies outside the main channel of airflow in a bypass. Getting the sensing element into a by-pass tube instead of having it in the main channel of airflow keeps the sensing element from impeding the flow of air into the engine. Being out of the main channel of airflow also reduces the build-up on the sensing element wire, eliminating the need for a burn-off cycle.

The Hitachi-style MAF sensor mounts directly on the throttle body assembly, which eliminates the possibility of a false air leak occurring between the MAF and the throttle.

Air Charge Temperature Sensor

The air charge temperature, or ACT, sensor is used to measure the temperature, and therefore the density, of the air in the intake manifold. Used only on the Port Fuel Injection cars, it measures the decrease in air charge density as the air moves through the warm intake manifold. This loss of density will result in less fuel being required for each cylinder because the expansion will cause each cylinder to be charged with less oxygen.

The air charge temperature sensor is a negative temperature coefficient thermistor like the coolant temperature sensor. It is fed a 5-volt reference, which is pulled low as the temperature drops and as the resistance of the thermistor drops.

Many Throttle Body applications use an air temperature sensor that measures the temperature of the incoming air before it passes the throttle plates and enters the intake manifold. As a result, even though Chevy calls it an air charge temperature sensor, the term is inaccurate. Operationally, the sensor used on the Throttle Body systems is the same as both the PFI ACT and the coolant temperature sensor.

Note: The temperature-resistance curves of the thermistors in the coolant temperature, air temperature and manifold temperature sensors are all identical.

Oxygen Sensor

The oxygen sensor might be described as a chemical generator. When it is heated to a minimum of 600 degrees Fahrenheit it will begin to produce a voltage ranging from 100 to 900 millivolts. Once operating temperature is reached, the sensor will begin to respond to changes in the content of exhaust oxygen. When the oxygen content of the exhaust is high, the ECM assumes that the engine is running lean. The design of the oxygen sensor is such that it will produce a low voltage when the exhaust oxygen content is high.

Oxygen content in the exhaust gases resulting from a lean combustion will result in a voltage less than 450 millivolts being delivered to the ECM. When the exhaust gases result from a rich combustion, the oxygen sensor voltage to the ECM will be greater than 450 millivolts.

When the oxygen sensor voltage is indicating a lean condition, the ECM will respond by enriching the mixture. When the oxygen sensor voltage is high, the ECM will respond by leaning out the mixture. In this manner the ECM adjusts for minor errors and variations from the rest of the input sensors, and controls the air-fuel ratio at 14.7:1.

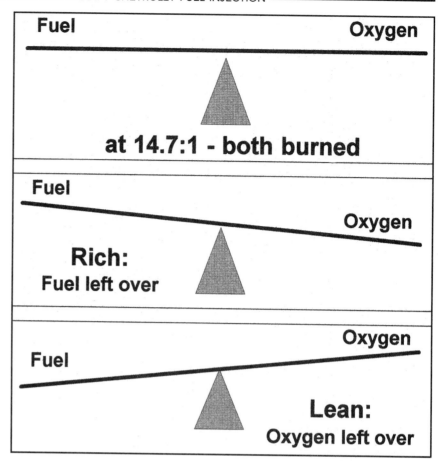

The theory behind the oxygen sensor is that an engine running at the correct air-fuel ratio of 14:7:1 will consume all of the oxygen and all of the fuel simultaneously. If the engine is running rich, then the oxygen will be consumed long before the fuel. The oxygen sensor has difficulty distinguishing between a rich-running and a perfect-running engine. However, a lean-running engine will leave large amounts of residual oxygen in the exhaust.

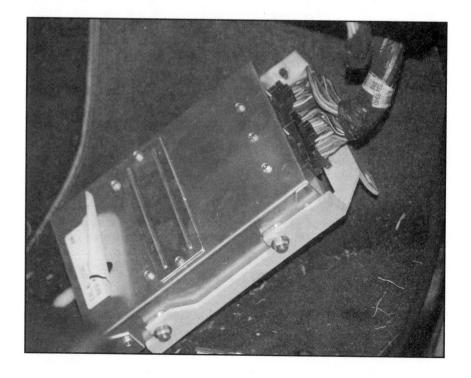

The ECM is the central processing and control unit for the fuel-injection system's sensors and actuators. Although this rather complex device is often blamed for hard-to-find drivability problems, it is almost never at fault. This is the ECM from a 1987 Camaro Z28. The small panel with two screws through it is where the PROM or CALPAK is located. This is also where the performance chips are installed.

Tach Signal

The tach signal is used by the ECM to synchronize, and sometimes sequence, the injectors. This signal always originates with some component in the primary ignition system, such as the ignition module or pickup. In addition to synchronizing the injectors, the ECM is designed to enter an enrichment mode at very high engine rpm.

Throttle Position Sensor

The throttle position sensor, or TPS, is a three-wire variable-voltage potentiometer sensor. A 5-volt reference is sent to the TPS by the ECM. The TPS is then grounded through the ECM, and a third wire carries the variable voltage that supplies the ECM with throttle position information.

The TPS is used by the ECM to make decisions concerning how much load is in the engine and what the driver expects the engine to be doing. TPS information is used in the control of the air pump, lock-up converter, acceleration air-fuel ratio control, and so on.

Electronic Control Module

The Electronic Control Module, or ECM, is the brain of the fuel injection system and is divided into three major sections: The ROM, the PROM and the RAM.

The ROM, or Read Only Memory, section of the ECM contains the principal set of instructions for the onboard computer to follow. This is the section that says, "When I see this happen, I have to make that happen." The microprocessor that contains these ROM instructions is a nonvolatile chip in which the programming designed into it cannot be erased by disconnecting the power.

The throttle position sensor measures the rotary angle of the throttle plates. The voltage delivered from the TPS to the ECM when the throttle is closed is about 0.5 volts. As the throttle is opened, the voltage should gradually and evenly increase to over 4.0 volts. Use a voltmeter or a scanner to test.

In 1994, Chevrolet expanded the ECM's control to include the transmission and other components and changed its name to Powertrain Control Module (PCM) to reflect the units more vital role. Aside from circuitry and programming to control other components, the only other significant difference between the older ECM and the new PCM is that PCMs do not feature a replaceable PROM chip. Instead, characteristics and parameters for a particular car are downloaded to an EEPROM in the PCM. Aftermarket companies, like HyperTech, already market performance-calibrated PCM "reprogrammers." *Jason Scott*

The PROM, or Programmable Read Only Memory, section is the calibration chip in the ECM. The PROM works along with the ROM to fine-tune the functions of fuel and timing control for the specific application. The PROM is also nonvolatile memory. It contains information about engine size, type of transmission, size and weight of the car, rolling resistance, drag coefficient, and final-drive ratio.

The PROM is serviceable separately from the ECM. In later applications it is serviced in a package along with a chip known as a MEM-CAL. This package of two chips is known as the CALPAK. Several aftermarket manufacturers such as Hypertech of Memphis, Tennessee (see appendices for address), produce replacement PROM chips that offer improved performance. Laws concerning the use of these chips differ from one state to another; in general, they are legal only for off-road use. The effect of these chips on the fuel injection and timing control system is similar to the effect of recurving the distributor and rejetting the carburetor in the old days.

The RAM, or Random Access Memory, section has three primary functions in the ECM. The first function is to act as the ECM's scratch pad; whenever a mathematical calculation needs to be done, the ECM uses the RAM. The second function is to store Block Learn Multiplier (BLM) information when the engine is shut down or operating in open loop. The third function is to store diagnostic codes when a system fault has been detected. These codes are stored for fifty restarts of the engine or until battery power is removed from the ECM (see chapter 10 on troubleshooting for more details), which means that unlike the ROM and PROM, the RAM chips are volatile memories.

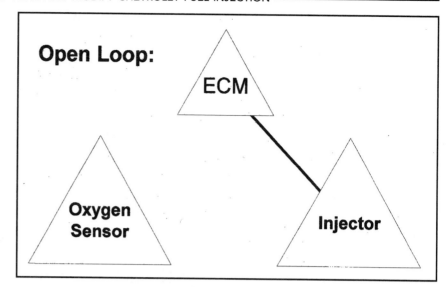

During the open-loop mode the ECM is making decisions about how long to leave the injector(s) open without information from the oxygen sensor. The fuel injection system will be operating in this mode whenever it believes that the engine or the oxygen sensor is too cold to operate in the more fuel-efficient closed loop mode.

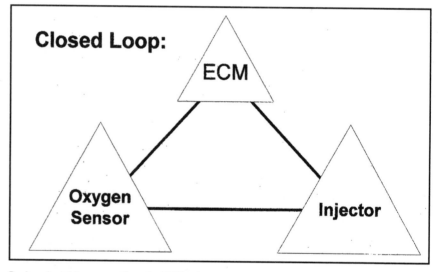

During closed-loop operation, the ECM selects an injector on-time based on information from the MAP, TPS, rpm and coolant sensor. After combustion, the oxygen sensor measures the oxygen content of the exhaust and reports back to the ECM on whether or not the quantity of fuel it injected was correct. If the oxygen sensor reports that the oxygen content of the exhaust was high, the ECM will assume that the engine was running lean and increase the injector on-time. If the oxygen content of the exhaust was extremely low, then the ECM will respond by shortening the injector on-time.

ECM Modes of Operation

General Motors fuel injection has eight distinct modes of operation: shutdown; start-up; open loop; closed loop; enrichment; enleanment; fuel cutoff; and clear flood. Each mode is described in the following pages.

Shutdown

The ECM enters the shutdown mode when the ignition switch is shut off for more than 2 seconds. For the first 2 seconds after the engine is shut off, the ECM remains powered up and the fuel pump remains activated. After the 2 seconds pass, all power is removed from the ECM circuits with two exceptions.

The first exception is the Diagnostic RAM. Trouble codes are retained during the shutdown mode for fifty restarts of the engine. The other exception is the BLM or Block Learn Multiplier RAM. The BLM remains powered and its memory retained so that when the engine is started cold, the ECM can compensate for the vacuum leaks and other air-fuel ratio problems it detected during its last operation in the closed loop mode.

Start-Up

The start-up mode is entered while the engine is being cranked. Timing becomes locked in at initial start-up and the air-fuel ratio is enriched slightly to expedite starting. This mode of operation continues for a second or two after starting to ensure that the engine will continue to run.

Open Loop

After the engine is started, it will operate in the open loop mode for several seconds to several minutes, depending on the temperature of the engine and oxygen sensor. In open loop, all engine sensors are actively providing information about running requirements, except for the oxygen sensor.

The ECM is watching the coolant sensor, the oxygen sensor and its internal timer to determine when the engine will be ready to operate in closed loop. Three things must occur before the ECM will enter closed loop. First, the coolant temperature sensor must reach a temperature specified for closed-loop operation by the PROM. Second, the ECM must know that the oxygen sensor is warmed up, which it does by monitoring its output voltage. When the voltage increases about 450 millivolts, drops below and then increases again, it is assumed that the oxygen sensor is warmed up. And third, a certain amount of programmed time must pass from start-up.

During open-loop operation, the coolant temperature sensor is telling the ECM about leaner requirements in the air-fuel ratio as the engine warms up, performing much the same function that a choke performs on a carbureted engine. The MAP or MAF is telling the ECM the quantity of air that is entering the engine, and about any changes in engine load.

The air temperature sensor, whether it be the one in the MAF or the one in the intake manifold, is updating the ECM on changes in fuel requirements based on the density of the air. The throttle position sensor is keeping the ECM abreast of what power changes the driver expects. As the TPS voltage increases, the ECM will assume it is because the throttle is opening. The computer will respond by increasing the amount of fuel that is entering the engine.

Closed Loop

The only real difference between open-loop and closed-loop operation is that during closed loop, the oxygen sensor is being used by the ECM to monitor exhaust oxygen content so that the air-fuel ratio can be constantly trimmed to 14.7:1. It is during closed-loop operation that the Integrator (or INT) and the Block Learn Multiplier are reprogrammed. Under normal driving conditions the ECM is in the closed-loop mode the majority of the time.

Enrichment

The fifth mode of operation is called the enrichment mode. The ECM can enter the enrichment mode from either open or closed loop. When entering this mode from closed loop the ECM will ignore the oxygen sensor until the engine operating conditions will once again permit closed-loop operation. Enrichment mode is entered when any one of the following conditions are met:

- *Wide-open throttle*—When throttle position sensor voltage goes over about 80 percent or 4 volts, the air-fuel ratio is enriched to provide extra power.
- *High engine load*—When the MAP sensor voltage approaches 4 volts, the ECM will enter the enrichment mode assuming that the engine is under a heavy load.
- *High engine rpm*—At extremely high engine rpm the air-fuel charge that is actually entering the combustion chamber tends to lean out. To maintain the proper 14.7:1 air-fuel ratio, the ECM enters the enrichment mode.
- *Overheating engine*—When the coolant temperature sensor detects an overheating engine, the enrichment mode is entered to cool the combustion chamber and thereby assist in cooling the operating temperature of the engine.

Enleanment

When the TPS indicates a nearly-closed throttle voltage, and

Open-Loop Operation

Monitors	Function
Coolant temperature sensor monitored	• Monitors engine temperature for entering the closed-loop mode • Monitors "choke" air-fuel ratio
MAF/MAP temperature sensor	• Monitors flow of air into the engine • Monitors engine load
Air charge temperature sensor Throttle position sensor	• Monitors changes in the density of the air • Monitors changes in driver demand and engine load based on driver demand

when the MAP or MAF indicates that there is little or no load on the engine, the ECM will lean out the air-fuel ratio to something leaner than 14.7:1. This serves two purposes. First, this is one means of conserving fuel to increase economy of operation. Second, the conditions that cause the ECM to enter the enleanment mode are also the engine operating conditions that cause high levels of carbon monoxide (CO) and hydrocarbon (HC) emissions. Leaning out the air-fuel ratio reduces the level of these emissions.

Fuel Cutoff

When the throttle is closed and the MAP or MAF indicates a negative load is on the engine, as in heavy deceleration, the ECM will cut off the injector(s). Again, CO and HC emissions are reduced as the engine rpm decreases. The ECM will remain in fuel cutoff during heavy deceleration until the engine rpm drops below about 1,500.

Clear Flood

Back in 1969, my dad taught me that when I thought my car was flooded, I should hold the accelerator pedal to the floor in order to maximize airflow and pray that the engine starts before the battery goes dead. Unfortunately, the natural response for the ECM is to notice the increase in TPS voltage, assume that the driver wants to accelerate and dump still more fuel through the injectors, flooding the engine even more.

To prevent this, the ECM has a Clear Flood program: When the engine is being cranked and the throttle is more than 80 percent toward wide open, the Clear Flood mode is entered, which either turns off the injector(s) completely or leans the air-fuel ratio to approximately 20:1.

As in the carburetor, the volume of air allowed to enter the engine is controlled by the throttle plates, also called throttle valves. As the volume of air entering the engine increases, the engine power, and therefore the engine speed, increases as well.

Contrary to popular belief, it is increased quantities of air (rather than increased quantities of fuel) that creates power. A dirty air filter can dramatically affect performance. One of the most common causes of low power in a fuel-injected car is a dirty air filter.

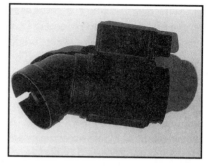

The Mass Air Flow (MAF) sensor must have a leak-free connection to the throttle assembly. Air that enters the induction system after the MAF is called "false air." Even a relatively small amount of false air can cause the engine to run lean, especially affecting performance at low engine speeds.

Although each of these modes of operation is a separate program within the ECM, they all act together to ensure good driveability, low emissions, and good fuel economy.

Air Induction System

Although we do not usually think of the air intake as part of the fuel injection system, its presence and proper operation is absolutely essential. We will break down this system into seven components: air cleaner; ducting; Mass Air Flow sensor (when used); throttle assembly; intake manifold; Idle Air Control valve; and cylinder head and intake valves.

Air Cleaner

A stock 350-ci Chevrolet engine can realistically flow about 405 cfm of air at 5,000 rpm assuming 8-percent volumetric efficiency. Flow bench tests by TPI Specialties show that the airflow potential of the stock air cleaner to throttle bore assembly is approximately 434 cfm for the Tuned Port 350-ci engine. This is more than adequate for the average commuter or enthusiast; however, a modified engine will require air cleaner modifications. Also, it should be noted that only 29 cfm of dirt restriction can occur even on a stock application before top-end performance begins to suffer.

Most Chevrolet fuel injection applications see daily use that does not require top-end performance, and the stock air cleaner is more than adequate. However, in chapter 6 on tuning, we will discuss modification and replacement of the air cleaner.

In addition to the air filter, the air cleaner housing may also contain the air charge temperature sensor.

Ducting

Ducting in the air induction system falls into two categories: Before measurement and after measurement. All of the ducting on TBI applications falls in the before-measurement category. The ducting on any PFI or TPI application that uses a MAF has before ducting in front of the MAF, and after ducting between the MAF and the throttle plates.

Problems in the before-measurement ducting will allow damaging airborne dust, dirt and sand to enter the engine. After measurement, ducting will not only admit contamination to the intake system but will also admit unmetered air known as "false air" into the intake manifold. False air tends to make the engine run lean and can create a driveability problem.

Although all of the ducting is a resistance to airflow, it is a necessary evil as the engine will be exposed to harmful dirt and debris without it. The ducting between the MAF and the throttle assembly on the PFI and TPI applications must be intact to prevent the entrance of false air into the intake system.

Mass Air Flow Sensor

Not only is the MAF part of the electronic measuring system on PFI or TPI, it is also part of the air induction system. This stock unit will permit the flow of 529 cfm of air for the 350-ci Tuned Port Injection applications. Again, this far exceeds what the engine is capable of sucking in. In chapter 11 on performance, you will find information about modifications that can raise the maximum cfm for the MAF all the way up to 750 cfm for use with trick engines.

Throttle Assembly

All components in the air induction system are passive except two, the Idle Air Control (IAC) valve and the throttle assembly. The throttle assembly consists of three major components: The Idle Air Control valve (which will be addressed as a separate component); the Throttle Position Sensor (TPS), which was discussed in the *Electronic Components* section of this chapter; and the throttle plates.

The throttle plates sit across the bore of the throttle assembly and control the flow of air into the intake manifold. The angle of the throttle plates is controlled by the driver through the throttle cable. The posi-

The ECM controls the Idle Air Control (IAC) valve on most Chevrolet fuel-injected cars to control the speed of the engine at idle. The two examples seen here are in their fully retracted positions. As the pintles extend, they decrease the amount of air entering the engine, slowing the idle speed. As they retract, they increase the amount of air, causing the idle speed to increase.

tion of the plates controls the volume of air entering the engine and therefore rpm.

Intake Manifold

The intake manifold transports the air from the throttle assembly to the cylinder head. On Chevrolet fuel-injected cars there are two types of intake manifolds used. The 5.0 and 5.7 liter TPI engines used a tunnel plenum with individual tube runners going to each of the intake ports in the cylinder heads. This allows for high air velocity with good low-end torque and high-speed horsepower.

All of the other Chevrolet manifolds are of a simple plenum design.

Idle Air Control Valve

The Idle Air Control, or IAC, valve is a stepper, motor-controlled valve which the ECM moves in order to control the speed of the engine at idle. The IAC can be moved to any 1 of 256 positions by the ECM to ensure the correct idle speed regardless of changes in engine load due to the transmission, power steering, alternator, air-conditioning compressor, or anything else.

At an idle, the IAC will be at about position 20 with no loads on the engine. As engine loads increase, the rpm will tend to drop. As the rpm drops, the ECM steps the IAC to a more open position (higher number);

as the rpm decreases, the IAC is stepped in. IAC position is displayed through serial data and is an important piece of troubleshooting information.

Cylinder Head and Intake Valves

Cylinder head and intake valves are covered in chapter 11 on performance modifications. However, the importance of the cylinder head and valves to the induction system cannot be overlooked. It is the opening of the intake valve as the piston begins to move downward on the intake stroke that creates the low-pressure area that allows ingestion of air.

Spark Control

In addition to controlling the fuel injection system, the ECM also controls ignition timing. The ignition timing is changed by the ECM in response to engine rpm, load and temperature.

Temperature

In the old days, when an engine was cold a thermal vacuum switch (TVS) was used to apply full manifold vacuum to the distributor vacuum advance, allowing for full advance and improved cold driveability. When the engine warmed up the TVS, also known as a coolant temperature override switch (CTO), would cut off the manifold vacuum and switch on ported vacuum to the vacuum advance. Chevrolet's computerized timing control system eliminates the CTO and uses the coolant temperature sensor input to the ECM in its place. When the engine is cold, the timing is advanced by the ECM to full advance; when the engine warms up, the idle timing is cut back and total advance is limited.

Engine Speed

The old-style distributors had a mechanical system for advancing ignition timing as the engine speed increased. The need to advance the ignition timing as rpm increases comes from the fact that combustion time is the same regardless of the

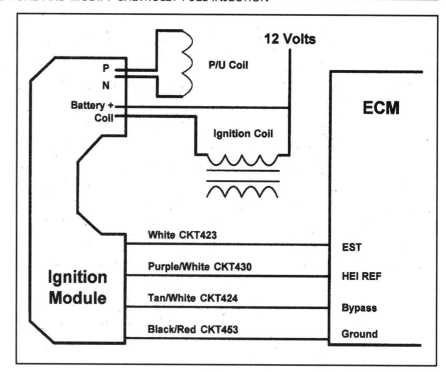

In addition to the fuel injection system, the ECM also controls ignition timing. Three major inputs provide information to the ECM for this function: coolant temperature, rpm and MAP (engine load). In many applications the presence of detonation is also detected with a knock sensor. The computer then controls the switch of the ignition module and coil with a variable duty cycle signal on the white wire (CKT 423).

speed of the engine—2 to 3 milliseconds. Yet, the higher the rpm, the more degrees of rotation the crankshaft covers in that time. The ECM on the late-model, fuel-injected cars receives a tach signal from the ignition module. As the engine speed increases, the ECM responds to this signal by advancing the timing to compensate for burn time.

Engine Load

When the engine is under a load, combustion pressure builds, increasing the possibility of detonation. Old technology addressed this problem by taking advantage of the fact that when the engine is under a load the manifold vacuum decreases. The decreasing manifold vacuum allows the vacuum advance unit to decrease the amount of advance, or in other words retard the timing when the engine load increases.

New technology uses the ECM's engine load measurement either directly from the MAP sensor for those cars equipped with one, or from the Load Variable from those not equipped with a MAP sensor, to retard the timing when the engine load increases. This accomplishes what the old technology did but with a greater degree of accuracy.

Detonation

Some Chevrolet Electronic Spark Timing (EST) systems incorporate a microphone-like device in the cylinder head or block called a Knock Sensor. The knock sensor detects vibrations that result from knocking or pinging. Back in the good old days, knocking and pinging were controlled by rich mixtures, leaded gas, and low compression ratios. Today's lean-running engines that run on unleaded gas at gradually increasing compression ratios have to depend on electronics to guard against detonation.

When a detonation is heard by the knock sensor, the timing retards 3 or 4 degrees each second until the

This solenoid-operated valve assembly controls and monitors vacuum applied to the EGR valve. Under the cap located on top of the solenoid there is a filter. If the filter becomes plugged, the EGR valve may be held open and the engine may die as the car is brought to a stop. This type of EGR control is found on the early 2.8-, the 5.0-, and the 5.7-liter engines.

knock goes away. It then begins to slowly advance the timing until—assuming that engine loads remain steady—the knock sensor begins to hear detonation. It backs up the timing slightly and will remain at that setting until the engine load or throttle position changes.

At first glance, a system to retard timing might seem to work against good performance and power; however, the opposite is true. A detonating engine is an engine that is wasting power. The air-fuel charge in the combustion chamber is being ignited too soon and is therefore trying to drive the piston back down the cylinder as the crankshaft is trying to push the piston up on the compression stroke. This robs power, wastes fuel and risks engine damage. The knock retard system controls detonation, thereby restoring the large part of this lost power and economy to the engine.

Exhaust Gas Recirculation Control

Several Chevrolet applications control the position of the Exhaust Gas Recirculation (EGR) valve with the ECM, and nearly all of them control whether or not the EGR will be allowed to open with the ECM.

The EGR valve has been considered a nuisance and power thief since its introduction in the early 1970s. It is true that the EGR valve renders useless approximately 7 percent of the combustion chamber volume for the creation of power, and therefore should be eliminated for track use where power and not emissions is the prime consideration. However, the average car owner will not feel a significant increase in performance or

When the 5.0- and 5.7-liter Tuned Port Injection engines are shut off, the ECM energizes a relay that sends a high current flow to the heated wire sensing element located in the MAF's main channel of airflow. This high current flow heats the element red-hot, burning off any road oil or contaminants that might be coating the sensing wire. This allows for accurate readings indefinitely through the life of the sensor.

economy when the EGR valve is eliminated. In fact, he or she may sense a loss of power. The EGR valve reduces nitrogen oxide emissions by using inert recirculated exhaust to cool the combustion process. This cooling reduces the tendency for the engine to detonate, and as previously discussed, detonation robs power.

Ported vacuum to the EGR valve is controlled by means of a vacuum solenoid which is controlled by the ECM. When the coolant temperature sensor detects a temperature below about 150 degrees Fahrenheit, the ECM closes the vacuum solenoid and the EGR valve does not function. When the temperature exceeds this PROM determined figure, the ECM grounds the vacuum solenoid, which then opens and permits the flow of ported vacuum to the EGR.

EGR Position Control

As stated, some applications also control EGR position with the ECM. For these, the control solenoid as described not only controls whether or not the EGR receives vacuum but also how much vacuum it receives. The amount of vacuum applied determines the position of the EGR valve.

2.8-Liter EGR

On the 2.8 liter engine the EGR control solenoid is a normally-closed solenoid. As long as there is no current flowing through the solenoid, the valve will remain closed and the EGR will receive no vacuum. When the ECM decides that conditions require the EGR valve to open, it grounds and un-grounds the solenoid about 10 times per second, varying the duration of each grounding—or in other words, varying the duty cycle to control the amount of vacuum going to the EGR valve and thereby controlling the EGR position. A duty cycle of 0 percent indicates that the EGR valve is being commanded by the ECM to remain closed. A duty cycle of 100 percent indicates that the ECM is commanding the EGR valve to open completely.

The air injection system (on applications that have one) pumps air into the intake manifold while the engine is warming up to assist in heating the oxygen sensor and catalytic converter. After the engine is warmed up (on many applications) the air is pumped into the rear, or oxidizing section of the converter, to assist in the conversion of hydrocarbons and carbon monoxide into carbon dioxide and water.

A vacuum diagnostic switch receives vacuum at the same time the EGR does. If the vacuum diagnostic switch does not cause the voltage at ECM terminal D8 to drop to zero when the EGR solenoid is commanded to deliver vacuum to the EGR valve, then the ECM will assume that there is either a defective solenoid or broken vacuum line.

5.0- and 5.7-Liter EGR

The 5.0- and 5.7-liter engines' EGR control system works exactly the opposite of the 2.8-liter's. The control solenoid has a normally-open valve that requires a 100-percent duty cycle to keep the EGR closed. As the duty cycle is decreased by the ECM, the vacuum delivered to the EGR increases and it opens more.

Another difference is the diagnostic switch. Both the 5.0- and the 5.7-liter engines use a temperature-sensitive switch located in the base of

the EGR or in the EGR tubing to detect when the EGR opens. As the EGR opens, the temperature of the base increases. This closes the temperature switch, causing the voltage at terminal C15 to drop from 12 to 0 volts. If the ECM does not detect this drop in voltage, then it will assume that the EGR did not open.

Stepper Motor EGR

Many of the newer-generation engines are using an EGR that is fully electronic. By making use of the same stepper motor technology as the Idle Air Control motor, the electronic EGR is able to accurately control the flow of exhaust gases into the combustion chamber without depending on engine vacuum. The stepper motor can be positioned far more accurately by using any hybrid of electronic and vacuum control. The result is far more dependable acceleration and deceleration characteristics.

MAF Burn-Off

The 5.0- and 5.7-liter Tuned Port Injection engines use a Bosch-style Mass Air Flow sensor. The Bosch MAF uses a heated wire to measure the amount of air entering the engine. This heated wire features a burn-off function that cleans the wire after the engine has been run by heating it red-hot.

Burn-off occurs when the engine is shut down after having been run in closed loop. The ECM grounds the black wire connected to terminal D12 which energizes the pull-down windings in the burn-off relay. This applies 12 volts to the heated wire, causing it to heat up, glow red, and burn off anything that may have stuck to it during the last drive.

Air Injection System

Computer control of the air pump system is accomplished (where applicable) through the use of two solenoid-operated valves known as the port solenoid and the converter solenoid. The port solenoid controls the pumping of air for upstream and downstream. When the engine is first started, the port solenoid is directing the air into the intake manifold upstream of the oxygen sensor. This helps to speed up the heating of the oxygen sensor and catalytic converter to operating temperature. As the engine approaches operating temperature the ECM un-grounds terminal A2 (on 5.0- and 5.7-liter applications), which de-energizes the port

solenoid and causes it to pump the air into the converter solenoid.

The converter solenoid controls the pumping of air between the catalytic converter and the atmosphere. Anyone who has ever owned or driven one of the smaller imports from the 1970s may have experienced a backfiring in the exhaust. In many cases, this backfire was caused by the air pump continuing to pump air into the exhaust during deceleration. During deceleration the content of CO (carbon monoxide) and HC (hydrocarbon) in the exhaust is high enough to be a combustible mixture. When air from the air pump contacts these exhaust gases, an explosion, or backfire, can occur in the exhaust. On PFI and TPI applications, the ECM controls the converter solenoid. When the ECM senses a negative engine load, a closed throttle and a decreasing rpm (as in deceleration), the ECM un-grounds terminal C2 and air is pumped to the atmosphere.

The default condition, which the Air Management System (AMS) will return to in the event of a component failure or a perceived fault in the fuel injection system, will be to pump air to the atmosphere.

Transmission Converter Clutch

Many automatic transmissions since the 1970s have featured lock-up torque converters. These lock-up converters allowed the engine and transmission to lock together in a one-to-one relationship, a very unusual

occurrence with a standard converter.

When the car reaches about 25 to 30 mph, with the transmission in high gear, a relatively light load on a warm engine and the throttle position sensor delivering a relatively steady signal to the ECM, the ECM will ground terminal A7 (this is the appropriate terminal for the Corvette and TPI Camaro; it may be another terminal on another application) and the torque converter clutch will lock up. Many manual transmission applications will use this same function to control an upshift light, which is used to signal that the driver should shift the transmission in order to maximize fuel economy.

Air Conditioner Compressor Clutch

Some applications control the air conditioner compressor clutch with the ECM. This assists in accurate control of the idle speed and allows the compressor clutch to be disengaged when the TPS approaches wide-open throttle, or when the engine is perceived to be under a heavy load.

Cooling Fan

Based on a signal from the coolant temperature sensor or the air conditioner compressor clutch, the ECM controls the cooling fan on some applications. ECM control of the radiator cooling fan may be used instead of or in addition to a temperature-sensitive fan switch.

C
O
N
T
E
N
T
S

FUEL 54

AIR 54

COMBUSTION 54

INCOMPLETE COMBUSTION 55

CARBON DIOXIDE 55

CARBON MONOXIDE 55

HYDROCARBONS 56

OXIDES OF NITROGEN 57

EMISSION CONTROL DEVICES 58

CHAPTER FIVE

AUTOMOTIVE EMISSIONS AND THE LAW

The act of combining air and fuel in the combustion chamber, raising the temperature of that charge through compression and igniting it with a spark plug, is a dirty and relatively inefficient way to move an automobile down the road. A plethora of both toxic and nontoxic chemicals are created that find their way into the atmosphere and the ground water. An understanding of the production and nature of these chemicals can assist in the understanding and troubleshooting of electronic fuel injection systems.

Fuel

Gasoline is a complex hydrocarbon made up of approximately 86 percent carbon and 14 percent hydrogen by weight. Trace impurities contribute to the noxious soup from the tailpipe. Sulfur can combine with oxygen during the combustion process, producing sulfuric acid and sulfur dioxide. There are also a wide range of additives that are used.

Until recently, tetraethyl or tetramethyl lead were used as knock inhibitors. When these oxidized in the combustion chamber they not only produced an acid but also contributed to the background lead contamination of the environment. New legislation, as well as the incompatibility of these chemicals with catalytic converters and oxygen sensors, has led to their disuse. Chemicals like methyl tertiary butyl ether (MTBE) are now used as knock inhibitors. Other additives such as anti-agers, detergents, anti-icers, and anticorro-sives also make up the soup you pour in your gas tank.

Air

As odd as it may sound, during combustion, air becomes one of its own prime defilers. About 78 percent of the air we breath is nitrogen. When this nitrogen is combined with oxygen during the combustion process, it can produce both nitrogen monoxide and nitrogen dioxide.

Air is made up of 78 percent nitrogen and 21 percent oxygen. The remaining 1 percent consists of various trace gases such as xenon, neon, and argon.

Combustion

When air and gasoline are mixed together in the combustion chamber, heated and then ignited, combustion occurs. For our purposes, combustion is the act of combining the oxygen in the atmosphere with the hydrogen and carbon elements of the gasoline. If we combine the air and fuel in the proper ratios (known as a stoichiometric ratio), apply just enough but not too much heat and then ignite the mixture with a sufficient spark, the perfect burn takes place. The carbon element of gasoline combines with the oxygen of the air to form carbon dioxide. The hydrogen element of gasoline combines with the oxygen element of the air to form water. The results of a scientifically mixed combustion of air and fuel at precisely the right burn temperature, therefore, is carbon dioxide and water.

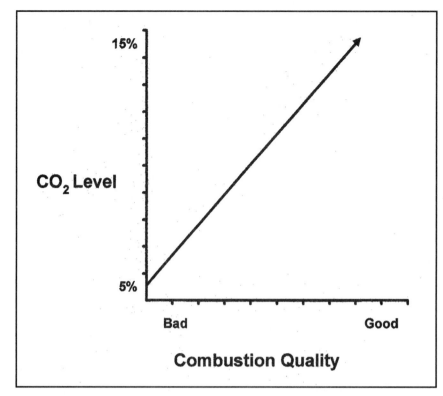

The process of combustion yields a soup of poisonous and harmful chemicals. Even with the best control of air-fuel ratio, timing and emission control systems, these chemicals will be present. What makes the limiting of automotive pollutants tricky is that the act of controlling one emission will often dramatically increase another. One of the desired by-products of combustion is carbon dioxide (CO_2). The better the quality of combustion, the higher the CO_2 level in the exhaust. Although there has been much recent controversy concerning the effects of CO_2 on the environment, there is no effort by the automotive industry to control CO_2. The fact is, virtually everything that is done to control emissions increases the output of CO_2.

Incomplete Combustion

Incomplete combustion will occur anytime one of the four elements of complete combustion gets out of balance: air, fuel, heat, or spark. Some of the toxic and unpleasant by-products of incomplete combustion or complete combustion that occur at too high a temperature include paraffins, olefins, aromatic hydrocarbons, aldehydes, keytones, carboxylic acids, polycyclic hydrocarbons, carbon monoxide, acetylene, ethylene, hydrogen, soot, nitrogen monoxide, nitrogen dioxide, organic peroxides, ozone, peroxyacetyl-nitrates, sulfur oxides, and from the additives, lead oxides and lead halogenides. It's a bit like finding out what they really put in sausage, isn't it?

We will address, define, and discuss only the by-products for which there is either a current EPA mandate or which can assist in fault diagnosis.

Carbon Dioxide

Currently, carbon dioxide (CO_2) is of no concern to the automobile industry. Listed as one of the results of complete combustion, the goal of most of the pollution control devices we have on the spark-ignition gasoline engine today is to increase the output of carbon dioxide. Yet carbon dioxide is not totally harmless. Increasing levels of atmospheric carbon dioxide has been linked to the Greenhouse Effect. It is theorized that the burning of fossil fuels as well as the depletion of worldwide vegetation are bringing levels to the point where solar heat will be trapped in the earth's atmosphere, increasing temperature aver-

ages world-wide. The automotive industry has not yet been mandated to lower carbon dioxide emissions, however.

Measuring carbon dioxide level at the tailpipe is a valuable diagnostic tool. A reading of between 10–15 percent indicates that the quality of combustion is good and there are no leaks in the exhaust system. The closer the reading is to 15 percent, the better the overall quality of combustion.

Low carbon dioxide levels in automobiles can result from the following: poor ignition quality (bad plugs, cap, rotor, plug wires); low compression (poor head gasket, valves, or piston rings); exhaust leaks (diluted sample); incorrect air-fuel ratio (too rich or too lean); and vacuum leaks.

Carbon Monoxide

Carbon monoxide (CO) is an odorless, colorless, tasteless gas that is highly toxic and dangerous. A 30-minute exposure to a CO concentration of only 0.3 percent by volume can be fatal. This percentage is well below even the California-mandated tailpipe reading. Even working on an emission-controlled vehicle with the engine running in a closed shop is dangerous. Additionally, your red blood cells have a 15-times greater affinity for carbon monoxide than they do for oxygen. As a result, long-term exposure to even lower levels can be health threatening.

Carbon monoxide is produced whenever the flame front runs out of oxygen as it travels through the combustion chamber. This oxygen deficiency can occur when there is not enough oxygen in the combustion chamber (such as from a restricted air filter), or when there is too much fuel (the engine is running rich). If the supply of oxygen had been sufficient, then the carbon element of the fuel would have been able to pick up two parts oxygen and complete its transition to carbon dioxide. Because of the inadequate supply of oxygen, the carbon could combine only with one part oxygen, stopping short at carbon monoxide.

Causes of high CO levels emitted from automobiles include a dirty air filter, high fuel pressure, leaking injectors (including cold-start), dirty oil and a tricked oxygen sensor.

For diagnostic and troubleshooting purposes, CO is used as an indicator of rich and lean running conditions. A CO level above 1 to 2 percent at the tailpipe on a catalytic-converter-equipped vehicle indicates that the engine must be running extremely rich.

Hydrocarbons

Hydrocarbon (HC) is a blanket term used for a wide range of toxic and carcinogenic chemicals produced during the combustion of gasoline. Many of these chemicals in the presence of nitrogen oxide and sunlight produce photochemical smog. This is the element of smog that burns the eyes and nose and irritates the mucous membranes. Although the environmentally conscious view hydrocarbon emission amounts as a veritable soup of foul substances, we will treat them as a single substance.

Allowable CO levels are relatively high; one to 2 parts per 100 (percent) is allowable even for late-model cars in most jurisdictions. The allowable concentration of various hydrocarbon emissions is extremely low, however, only about 100 to 300 parts per million. To compare, 1 percent equals 10,000 parts per million, or 1 part per million equals 0.000001 percent. In percentages, therefore, an allowable level of hydrocarbons is only about 0.002 to 0.003 percent.

Excessive hydrocarbon content in the exhaust can be caused by sev-

Oxides of nitrogen (NO_x) are among the hardest emissions to control. As the combustion temperature increases, the output of NO_x also increases. This is especially complicated by the fact that most things done to control hydrocarbon and carbon monoxide (CO) emissions tend to increase combustion temperature.

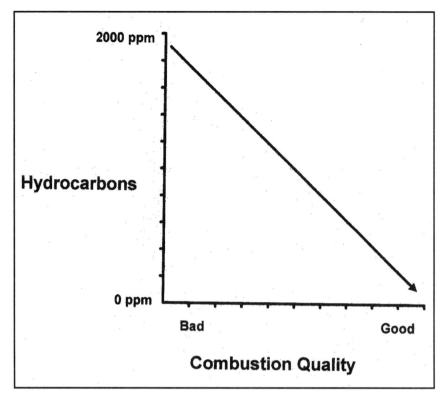

Hydrocarbon (HC) emissions are basically unburned fuel. These were among the first emissions that the automotive industry tried to clean up. As long as the quality of combustion—which is a function of air-fuel ratio, ignition and compression—remains good, then hydrocarbon emissions will remain low. As the quality of one of these three deteriorates, the quality of combustion deteriorates and the hydrocarbon emissions increase.

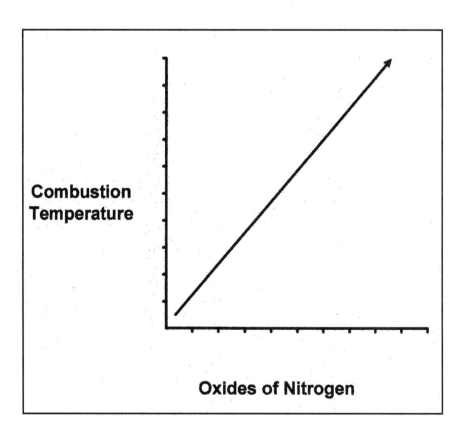

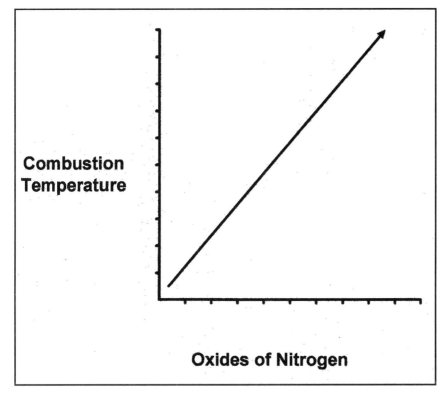

Combustion Temperature

Oxides of Nitrogen

Oxides of nitrogen (NO_x) are among the hardest emissions to control. As the combustion temperature increases, the output of NO_x also increases. This is especially complicated by the fact that most things done to control hydrocarbon and carbon monoxide (CO) emissions tend to increase combustion temperature.

eral factors including poor secondary ignition quality, vacuum leaks, engine running too rich, lean misfire (engine running too lean) and low compression.

Oxides of Nitrogen

Oxides of nitrogen include both nitrogen monoxide and nitrogen dioxide. Usually these two chemicals are grouped together and referred to as nitrous oxide (NO_x). Nitrogen monoxide is an odorless, colorless, tasteless gas that is basically harmless to the environment. However, created along with the NO is nitrogen dioxide. Nitrogen dioxide is a reddish brown, poisonous gas that destroys lung tissue. To further complicate matters, nitrogen monoxide combines with atmospheric oxygen to become nitrogen dioxide.

NO_x is created when combustion temperatures exceed 2,500 degrees Fahrenheit. At these temperatures the nitrogen combines chemically with oxygen (burns). Although NO_x is generally not measured for diagnostic purposes, it is the principle emission responsible for oxygen sensor-controlled fuel injection systems. The allowable level of NO_x since 1982 for new cars delivered in the United States has been 1.0 gram per mile.

Since there is currently no economical way to measure NO_x, it is of no diagnostic value. NO_x levels will increase, however, when the combustion temperatures increase as a result of factors such as incorrect initial timing, lean air-fuel ratio, high compression (such as caused by carbon buildup on the pistons) and vacuum leaks.

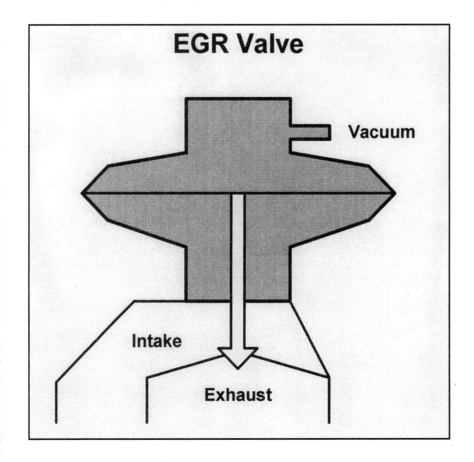

EGR Valve

Vacuum

Intake

Exhaust

The EGR valve meters relatively cool but inert exhaust gases into the intake manifold to reduce combustion temperatures. The precise control of the amount of EGR gases entering the intake is critical. If too much is allowed to enter, the combustion chamber temperature will drop to the point where the engine will misfire, stumble or stall. Not enough EGR and the engine will ping.

Emission Control Devices

Engine Modifications

By dropping the compression ratios in the mid-1970s, manufacturers were able to lower combustion temperatures and thereby lower the production of NOx. This caused fuel economy to plummet and totally destroyed the potential for performance.

Increased valve overlap allows for some of the exhaust gases to be drawn back in during the beginning of the intake stroke. These exhaust gases are inert; the fuel and the oxygen were burned out of them during the previous combustion cycle. As the next combustion cycle begins, these inert gases act as a heat sink to lower the combustion temperature and decrease the production of NOx.

Exhaust Gas Recirculation Valve

The exhaust gas recirculation, or EGR, valve is used to reduce the production of oxides of nitrogen. As the engine rpm rises above idle, either vacuum applied to the diaphragm or a solenoid lifts the pintle of the EGR valve, allowing exhaust gases into the combustion chamber. Approximately 7 percent of the combustion chamber volume will lower the burn temperature by about 500 degrees Fahrenheit. This lower temperature reduces the production of NOx, and also the likelihood of detonation occurring.

Positive Crankcase Ventilation

The positive crankcase ventilation, or PCV, system draws crankcase vapors through a metered check valve known as the PCV valve into the intake manifold. These gases, which are the result of combustion chamber blowby, are given a second chance to be burned. The PCV system is designed to reduce the amount of HC (hydrocarbons) allowed to escape to the atmosphere from the crankcase.

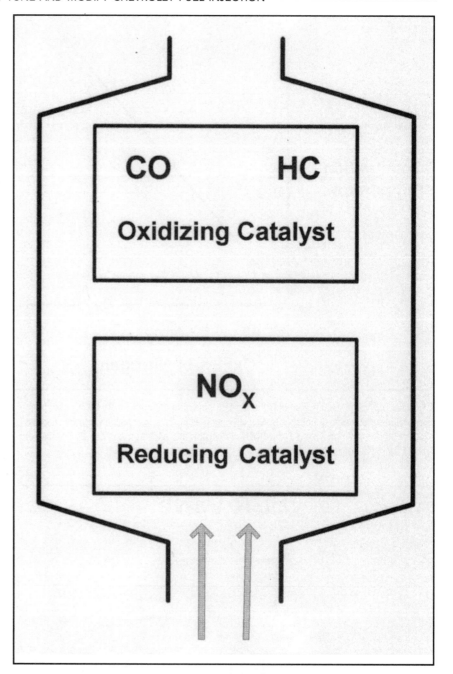

The catalytic converter has been blamed for power loss and poor running engines since its introduction in the mid-1970s. In reality it contributes no significant back pressure to the exhaust of a stock engine. The converter used with Chevrolet fuel injection is called a dual-bed converter, meaning that there are two separate catalyst elements. It is also sometimes referred to as a three-way converter, since it controls CO, HC, and NO_x.

Fuel Evaporation

Since the 1970s, cars have been equipped with an evaporative control system. This system consists of a canister filled with activated charcoal. The canister is then connected to the fuel tank (and in the case of carbureted cars, to the carburetor fuel bowl) by hoses. As gasoline (HC) evaporates from the fuel tank, the fumes are stored in the activated charcoal. A device known as the canister purge valve opens when the engine can accept the extra fuel, and the manifold vacuum sucks the fuel into the engine to be burned.

Air Pumps

The air pump has two purposes. First, it pumps air into the exhaust manifold while the engine is cold. Since a rich mixture is present in the combustion chamber at this time, the resulting exhaust gases will be laden with CO and HC. When the oxygen in the pumped-in air comes in contact with the CO and HC, their temperature increases dramatically and afterburning takes place. The afterburn consumes much of the residual CO and HC.

As a side benefit from this afterburning, a great deal of heat is generated. The heat is used to assist in bringing the catalytic converter and the oxygen sensor up to their proper operating temperature.

Then, before the ECM goes into its closed-loop mode, the air pump diverts the air either to the atmosphere or downstream of the oxygen sensor.

Catalytic Converter

When the catalytic converter was introduced in 1975, it spelled the beginning of the end for leaded gasoline. This oxidizing catalytic converter consisted of a platinum-palladium coating over an aluminum oxide substrate. The substrate over the years has been in the form of pellets, beads and a honeycomb monolith. The platinum was not compatible with tetraethyl lead, preventing its use in the fuel used in catalyst-equipped vehicles.

The job of the catalyst is to provide an environment where enough heat can be generated to allow further combustion of the HC and CO to occur. The converter is heated by a chemical reaction between the platinum and the exhaust gases. The minimum operating, or "light-off" temperature of the converter is 600 degrees Fahrenheit with an optimum operating temperature of about 1,200 to 1,400 degrees Fahrenheit. At a temperature of approximately 1,800 degrees Fahrenheit the substrate will begin to melt. These excessive temperatures can be reached when the engine runs too rich or is misfiring.

I once attended a meeting of the Society of Automotive Engineers (SAW) where a representative of a company that built the converter substrates stated that a 25-percent misfire (one cylinder on a four-cylinder engine) for 15 minutes was enough to begin an irreversible self-destruction process in the converter. This type of catalytic converter requires plenty of oxygen to do its job, meaning that the exhaust gases passing through it must be the result of an air-fuel ratio of 14.7:1 or leaner.

Between 1978 and 1982, we saw the gradual introduction of the dual-bed converter. This converter added a second rhodium catalyst known as the reducing section ahead of the oxidizing section. The rhodium coats an aluminum oxides substrate and reacts with the NOx passing through it. When heated to an excess of 600 degrees Fahrenheit, the nitrogen and oxygen element of the NOx passing through it will be stripped apart. Although only about 70- to 80-percent efficient when coupled with the EGR, it does a dramatic job of reducing NOx. Since the job of the reducing catalyst is to strip oxygen away from nitrogen, it works best when the exhaust gases passing through it are the result of an oxygen-poor air-fuel ratio of 14.7:1 or richer.

The only air-fuel ratio that will permit both sections of the converter to operate efficiently is a ratio of 14.7:1. The job of the oxygen-feedback fuel injection systems being used today is to precisely control the air-fuel ratio at 14.7:1 as often as possible.

Note: On many applications the air from the air pump, after completing its role of preheating the catalytic converter, will be directed between the front reducing section and the rear oxidizing section of the converter to supply extra oxygen, thus improving the efficiency of the oxidizer.

SPARK PLUGS	**60**
COMPRESSION	**61**
AIR-FUEL RATIO	**61**
SPARK PLUG WEAR	**61**
HIGH-VOLTAGE WIRES	**61**
SPARK PLUG REPLACEMENT	**62**
SPARK PLUG INSTALLATION	**62**
SPARK PLUG WIRE TESTING AND REPLACEMENT	**63**
DISTRIBUTOR CAP REPLACEMENT	**63**
AIR FILTER REPLACEMENT	**64**
FUEL FILTER REPLACEMENT	**64**
ALCOHOL CONTAMINATION	**64**
TESTING IGNITION TIMING	**65**
ELECTRONIC SPARK TIMING SYSTEM TESTING	**66**
FAILURE OF THE EST TO ADVANCE TIMING	**66**
SPARK TIMING-OUT WIRE TESTING	**66**
EST BYPASS WIRE TESTING	**66**
AIR-FUEL RATIO TESTING	**67**
OXYGEN SENSOR TESTING	**69**
MINIMUM AIRFLOW	**69**
CLEANING COKING	**70**
MINIMUM AIRFLOW ADJUSTMENT	**70**
PFI AND TPI, MODELS 220 AND 700	**70**
TBI MODELS 300 AND 500	**71**
CROSSFIRE INJECTION	**71**
THROTTLE BODY INJECTOR CLEANING	**71**
DISASSEMBLY	**71**
REASSEMBLY	**71**
PORT AND TUNED PORT INJECTOR CLEANING	**71**
DEAD CYLINDER TESTING: DISTRIBUTOR IGNITION	**72**
DEAD CYLINDER TESTING: DISTRIBUTORLESS IGNITION	**72**
TUNING SERIAL DATA	**74**
TUNING PROCEDURE	**75**
THE 100,000-MILE TUNE-UP	**75**

CHAPTER SIX

FUEL INJECTION TUNING

Fuel injection, computerized engine controls, tough emission standards, and extended service intervals have not eliminated the need for regular maintenance and routine tune-ups. Many cars of the late 1980s and into the 1990s boast of spark plug replacement at intervals of 50,000 miles or more. There are still air filters and fuel filters with similar service intervals. These service intervals may be fine for the typical consumer who uses his or her car to go to and from work, with an annual trip to Yellowstone or Disneyland. However, the owner who expects a little more from a car—more performance, more economy, or more reliability—should consider the following tune-up maintenance schedule.

Every 12,000 Miles
Replace spark plugs
Replace fuel filter
Replace air filter
Check vacuum hoses and intake air tubes
Scan for codes and voltages

Every 24,000 Miles
Replace spark plugs
Replace fuel filter
Replace distributor cap
Replace rotor
Replace plug wires
Check vacuum hoses and intake air tubes

Spark Plugs

Replacing the spark plugs is one of those tasks that gives every Saturday afternoon mechanic a feeling of pride. Yet there are many subtle details to learn about spark plug replacement, and about what the old spark plugs can tell you. Let us begin with some of the basics of what the spark plug does.

The spark plug ignites the air-fuel charge in the combustion chamber. It accomplishes this by means of a high-voltage spark across its electrodes. The source of the spark is the ignition coil. The spark is conducted to the spark plugs by means of the secondary ignition wires. This spark lasts for about 1.5 to 2.0 milliseconds at around 1,000 volts. However, for the first 30 microseconds the voltage of the spark is considerably higher, somewhere between 5,000 and 30,000 volts. The higher voltage is necessary to initiate, or begin, the spark across the gap of the plug.

Since the ignition coil is basically a transformer, the amount of energy available to the plugs is limited to the number of watts passing through the primary. If too much of the available energy is used to initiate the spark, then the amount of energy left to maintain the spark will be reduced. Today's leaner-running engines tend to extinguish the fire in the cylinder when the spark goes out. Therefore, anything that affects the duration of the spark affects power and driveability a lot more than it did in the 1960s. Anything that affects the voltage required to initiate the spark will affect the duration of the spark.

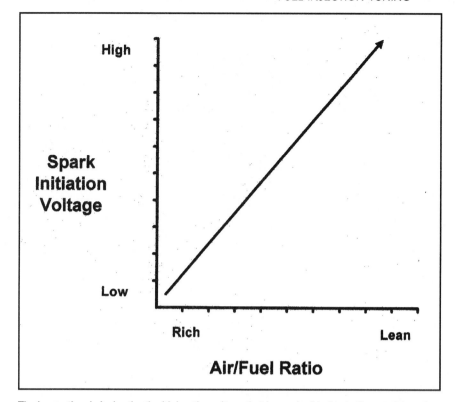

The leaner the air-fuel ratio, the higher the voltage that is required to begin the spark jumping the gap across the spark plug. The more energy that is used to initiate the spark, the less energy there is to maintain the spark across the gap. As the engine runs progressively leaner, the likelihood of misfire increases. And as the spark plug wears, the likelihood of misfire increases again.

Spark Initiation Voltage

A number of factors affect the spark initiation voltage, including spark plug heat range, compression (and compression ratio), air-fuel ratio, spark plug wear, and condition of distributor cap, ignition rotor, ignition coil, and secondary ignition wires. Each factor is covered in detail in the following pages.

Spark Plug Heat Range

Unless you have a highly modified engine, you should stick with the heat range recommended by Chevrolet or the spark plug manufacturer. If the incorrect heat range selected is too cold, it can greatly increase the spark initiation voltage, robbing power from the engine. Another difficulty with a cold plug is that it can result in misfiring and fouling. Using spark plugs with too hot of a heat range can result in pre-ignition or pinging.

Compression

The effect of compression on the power from an engine is like a two-edged sword. On the one hand, increased compression greatly increases the power potential of the engine; yet increased compression also increases spark initiation voltage. The increase in spark initiation voltage means that high-performance, high-compression engines may require an ignition coil with greater potential energy. However, as a stock engine's compression in each cylinder drops, the voltage required to initiate the spark decreases—but so does engine power.

Air-Fuel Ratio

Air-fuel ratio is one of the most critical variables when it comes to driveability. If the air-fuel ratio is too rich, spark initiation voltage will be low and the burn in the combustion chamber will be slow and incomplete. A lean air-fuel ratio makes it difficult for the spark to start jumping the spark plug gap. A great deal of energy will be used during the first 30 microseconds, leaving little energy to maintain a burn in the combustion chamber. The result is reduced power, stumbling, hesitation, and misfiring.

Spark Plug Wear

If you have ever removed an old spark plug, you may have noticed that the gap is much wider than it was when it was installed. This is because, like many other components in the engine, spark plugs wear as they are used. The wider the gap the higher the voltage required to begin the spark jumping the gap, and again the less energy there is to accomplish the burn in the combustion chamber. This has a direct effect on combustion, decreasing power, increasing toxic emissions, and causing poor driveability.

High-Voltage Wires

Within the high-voltage side of the ignition system there are several other items that wear as they work. The distributor cap, on cars equipped with a distributor, is used along with the rotor to transfer the high-voltage spark from the ignition coil to the spark plug wires. High voltage arrives at the center of the distributor cap either through the coil wire or directly from the ignition coil. This voltage is conducted to the rotor by means of a carbon nib. The current then travels through either a solid metal conductor or a resistive element in the rotor. As the rotor swings past each of the plug wire electrodes, a spark jumps from the rotor.

Wear on the carbon nib or an excessive gap between the rotor and the distributor cap electrodes will increase the amount of energy consumed on the way to the spark plugs, thus decreasing the amount of energy available to fire the cylinder. Distributorless ignition systems are one of the methods that Chevrolet is using to reduce the number of places where wear can affect secondary ignition energy, thereby increasing tune-up service intervals.

As plug wires age, their resistance increases and as their resistance increases, a greater portion of the potential spark energy heading toward the spark plugs is consumed.

Spark Plug Replacement

With the engine cold, unscrew the spark plugs two or three turns. Using compressed air or a solvent-soaked brush, clean the area around each plug to remove any dirt or foreign objects that might fall into the cylinder when the spark plug is removed. Should you find the spark plug difficult to unscrew, it might be that the plug has dirt or grit on the threads. If this occurs, put a couple of drops of oil on the exposed threads and allow it to soak in for a few minutes. Screw the plug back in, then out several times, a little more each time until the plug is removed.

Inspection of the removed spark plugs can tell you a lot about the running condition of the engine. Carbon deposits, ash formations, oil fouling, sooting and other conditions can indicate engine or injection system problems.

Spark Plug Installation

Before installing the new spark plugs be sure to check the gap. Although some spark plug manufacturers make an effort to pre-gap their plugs, they can be unintentionally "regapped" in shipping. Inspect the threads in the cylinder head to ensure that they are clean and undamaged. Also check the mating surface where the plug contacts the head. It should be clean and free of burrs.

Start the new spark plug and screw it in a few turns by hand. Continue to screw in the plug either by hand or with a socket and ratchet until it contacts the mating surface firmly. To avoid over-tightening, use a torque wrench and tighten a tapered seat plug to 26 lb-ft in a cast-iron head, or 21 lb-ft in an aluminum head. In the real world a torque wrench is seldom used when

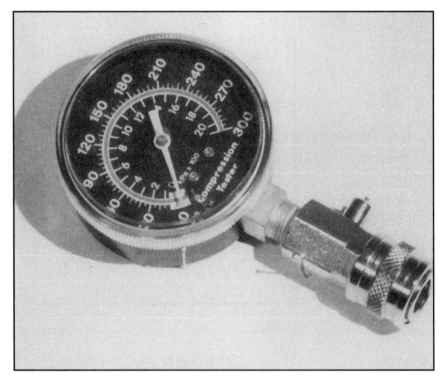

When a drivability problem on a fuel-injected car is being traced, compression should be tested early. If the engine, valvetrain, rings, and so on are in poor shape, then the fuel injection system will not behave properly. Always begin with the basics.

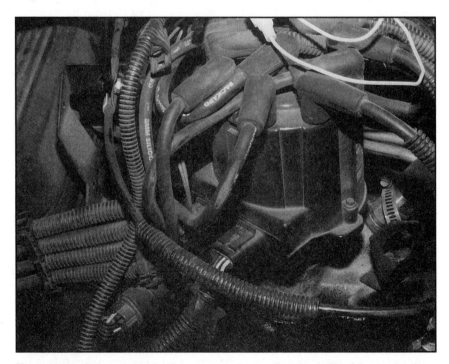

The condition of all secondary ignition components can affect the quality of the combustion process. Inspect not only the spark plugs but also the condition of the plug wires, the distributor cap and the rotor.

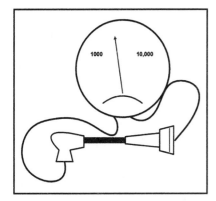

The ohmmeter remains the best tool for testing the spark plug wires. Connect an ohmmeter to each end of the plug wire. The resistance should be between 1,000 and 10,000 ohms per foot.

The air filter may need to be replaced more often than the recommended service interval, depending on the environment where the vehicle is being operated. Not only does a dirty air filter restrict the flow of air into the engine, but it also may allow dirt and contamination to enter the engine, damaging cylinder walls, pistons, and valves.

installing spark plugs, so use this rule of thumb: firm contact +1/5 rotation.

Spark Plug Wire Testing and Replacement

There are two ways of testing spark plug wires. The first is with the use of an engine analyzer oscilloscope. The second method is with the use of an ohmmeter. Remove each plug wire, and with your ohmmeter on the X1000 ohm scale measure their resistance end to end. A good plug wire will have less than 10,000 ohms but greater than 1,000 ohms per foot.

There are only a couple of important things to remember when replacing secondary ignition wires. First, if the plug wires are installed in the incorrect order a backfire may occur, resulting in damage to the Mass Air Flow sensor or the rubber tube that connects it to the throttle assembly. The original-equipment plug wires have numbers on them indicating to which cylinder they should be connected. Aftermarket or replacement plug wires may not have these numbers. 3M and other companies make adhesive numbers that can be attached to ensure proper reinstallation.

Distributor Cap Replacement

As early as 1974, Chevrolet began using their high-energy igni-

tion system with the coil-in-the cap distributor. Several V-6 and V-8 fuel-injected applications still made use of this distributor through the latter half of the 1980s.

When replacing the distributor cap, the coil must be removed from the old cap (unless it is also being replaced) and installed in the new cap. Take care that the three little components supplied with the new cap—the rubber washer, the spring and the carbon nib—are installed in the proper sequence.

The rotor on these distributors is installed with a pair of screws. A square and a round fail-safe stud prevent installing it backwards.

The remote-coil distributor cap is more conventional in design than the cap previously mentioned. Replacement consists of simply releasing the attachment screws or clips and making sure that the plug wires are reinstalled in the correct order.

Whichever type of distributor you have on your car, it would be a good idea to replace the distributor

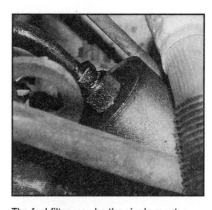

The fuel filter may be the single most important service item in the fuel injection system. Service intervals are often written with an optimistic opinion of fuel quality. Replacing the fuel filter every 24,000 miles is a sensible interval.

cap and rotor together and use the same brand. Pairing caps and rotors of two different manufacturers can result in the incorrect rotor air gap. Excessive rotor air gap can cause excessively high spark initiation voltage and can result in incomplete combustion.

Air Filter Replacement

The real value of the air filter is significantly underestimated. It is the engine's only defense against sand, grit and other hard particle contamination. When these substances enter the combustion chamber they can act like grinding compound on the cylinder walls, piston rings, and valves. Replace the air filter at least once a year or every 24,000 miles. In areas where sand blows around a lot, like West Texas or Arizona, the air filter should be replaced much more often.

On most carbureted cars, a restricted air filter will cause the engine to run rich. This is because the restriction causes a reduction of pressure in the ventures of the carburetor while the pressure in the fuel bowl remains constant at atmospheric pressure. This increased pressure differential increases the flow of fuel into the ventures and the mixture is enriched.

On a Port Fuel Injection engine, the ECM measures the exact volume of air entering the engine by the Mass Air Flow sensor, and meters in the correct amount of fuel for the measured amount of air. Restrict the flow of incoming air and less air will be measured, so less fuel will be metered into the engine.

Although Throttle Body Injection makes use of a different air measurement technology, it does much the same thing as Port Fuel Injection. TBI air measurement is a little less precise and therefore may run a little rich as a result of a restricted air filter. Even on the TBI cars, the oxygen sensor should compensate for any air-fuel ratio errors that may occur once closed-loop operation begins.

Fuel Filter Replacement

The fuel filter is the most important service item among the fuel components of the fuel injection system. In my 15 years experience as a technician on fuel-injected cars, I have replaced many original filters on cars that were over 10 years old. This lack of routine maintenance is just begging for trouble.

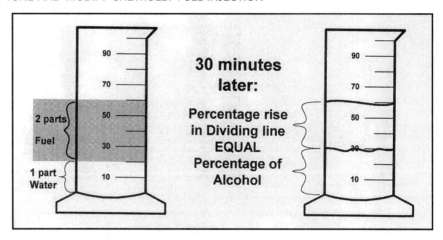

Alcohol contamination can seriously damage fuel-injection components. If you have an expensive component that has suffered failure due to fuel damage, put one part water and two parts of the fuel in question into a graduated cylinder. Note the dividing line. Allow the cylinder to sit undisturbed for 30 minutes. The percentage rise in the dividing line is the percentage of alcohol in the fuel. No more than 10 percent is advisable.

Begin by removing the fuel filter. Find a white ceramic container, such as an old coffee cup, and drain the contents of the filter through the inlet fitting. Inspect the gasoline in the cup for evidence of sand, rust or other hard particle contamination. Now pour the gas into a clear container such as an old glass and allow it to sit for about 30 minutes. If there is a high water content in the fuel, it will separate while sitting; the fuel will float on top of the water. If there is excessive water or hard particle contamination in the tank, it may have to be removed and professionally cleaned. For minor water contamination problems there are a number of additives that can be purchased at your local auto parts store.

Should the fuel filter become excessively clogged, the following symptom might develop: You start the car in the morning and it runs fine. As you drive several miles down the road, the car may begin to buck a little or lose a little power; then suddenly the engine quits almost as though someone had shut off the key. After sitting on the side of the road for several minutes, the car can be restarted and driven for a couple of miles before the symptom recurs.

This problem could be caused by a severely restricted fuel filter. The car runs well initially because

the bulk of what is causing the restriction has fallen to the bottom of the fuel filter as sediment. When the engine is started and the fuel begins to flow through the injection system, this sediment gets stirred up and pressed against the paper elements of the fuel filter. As it does, fuel volume to the injectors is decreased and the engine begins to run lean. Sooner or later the engine leans out so much that it dies. But the really bad news is that often when the fuel filter becomes that restricted, some of the contaminants have forced their way through the filter and may contaminate or damage the rest of the components in the fuel system.

Alcohol Contamination

Alcohol contamination can damage many of the fuel injection components. If you suspect that excessive alcohol content has caused the failure of system components, then you might want to test a fuel sample for alcohol content.

A rather simple test has been recommended by GM and others. Pour 200 milliliters (ml) of the sample fuel into a glass or clear plastic container along with 100 ml of water. Immediately after putting the two liquids in the container, the dividing line will be at the 100 ml mark. Wait about 30 minutes; if the dividing line

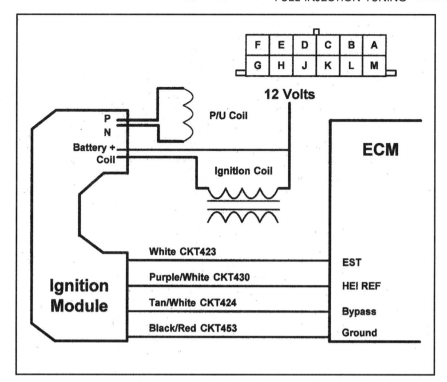

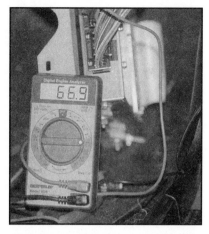

The white wire that runs between the ECM and the ignition module carries the timing control signal. Connect a dwell meter set on the four-cylinder scale to the white wire. As the engine rpm or load changes, the dwell should also change. This would indicate that the ECM is attempting to control the ignition timing.

Placing a jumper between terminals A and B of the ALDL connector should lock the timing control by the ECM so that initial timing can be checked. A more definite way of locking the timing is to disconnect the tan and black wires that run from the ECM to the ignition module. Although an inline connector is provided for this purpose, it is often difficult to find.

rises by more than 10 percent of the volume of the contents of the container, then there is excessive alcohol in the fuel. Drain the fuel from the tank and replace it with good fuel.

There are many fuel additives on the market that contain alcohol. The alcohol content in most additives is so small, however, compared to the size of the typical fuel tank that it poses no threat to the fuel system. Nevertheless, use caution, ask around, and be selective when purchasing these products. Some are much better than others.

Testing Ignition Timing

On 1982 and later fuel-injected Chevrolets, the ignition timing is controlled by the fuel injection computer, the Electronic Control Module (ECM). There are four wires involved in the control of ignition timing. All of them run from the distributor to the ECM. A purple wire with a white tracer sends a tach signal from the ignition module to the ECM. The ECM then modifies the signal by changing its duty cycle and returns it to the ignition module through a white wire. In the ignition module this signal is used to control ignition timing.

There is also a tan and black wire that signals the ignition module to ignore the signal on the white wire while the engine is being cranked. During cranking there is 0 volts on the tan and black wire; after the engine starts, the ECM puts 5 volts on the wire, which notifies the ignition module to use the signal on the white wire to control ignition timing. The fourth wire is a black wire that usually has a red tracer; this wire is simply a common ground between the ECM and the ignition module.

To check the ignition timing, follow the instructions on the EPA decal found under the hood. The following instructions may vary a bit from those for your vehicle. First, disconnect the tan and black wire located somewhere in the engine compartment.

This will turn on the Check Engine light and set a trouble code number 42 as soon as the engine is started. Start the engine, adjust the rpm according to the specification on the EPA decal, check or set the initial timing, then reconnect the tan and black wire and clear the code.

Place a jumper wire between terminals A and B of the ALCL connector. Check or adjust the timing to manufacturers specifications. This method will set no code.

Disconnect the four-wire connector located near the distributor. Adjust timing as described. This will also set a Code 42; do not forget to clear it when you are done.

After preparing to check initial timing through one of the methods already described, use a timing light in the normal fashion to check the timing. Adjustment of the timing is still done through loosening and rotating the distributor. On some applications the timing advance system will not begin to function again until the engine has been shut off and restarted.

If you are working on a car equipped with a distributorless ignition system, then you will not find an initial timing specification or adjustment procedure. As discussed later in this book, on these ignition systems the timing is controlled entirely by the computer.

Electronic Spark Timing System Testing

With the car in park and the parking brake set, rev the engine while monitoring the timing with a timing light. The timing should advance considerably. At this point you are not so much concerned with how much it advances as you are with the fact that it does advance. Distributorless ignition systems can be tested for timing advance by making your own marks on the harmonic damper and the timing chain cover. Even though there is no adjustment for initial timing, proper advance of timing is still essential for good acceleration and power. If the timing does not advance, the possibilities are as follows:

- You did not cycle the ignition key off, then restart after checking initial timing.
- Some applications require that the transmission be in forward or reverse gear before timing is advanced by the ECM. Timing advance on these can only be checked when the drive wheels are elevated or on a chassis dynamometer.
- There are even some applications that require an indicated vehicle speed before timing advance can occur. Again, timing advance on these cars can be tested only with the drive wheels elevated or on a chassis dyno.

Failure of the EST to Advance Timing

If you have checked the possibilities mentioned and the timing still does not change in response to changes in rpm, then it is time to grab the voltmeter and wiring diagram.

Spark Timing-Out Wire Testing

There is a white wire referred to as the spark timing-out wire that the

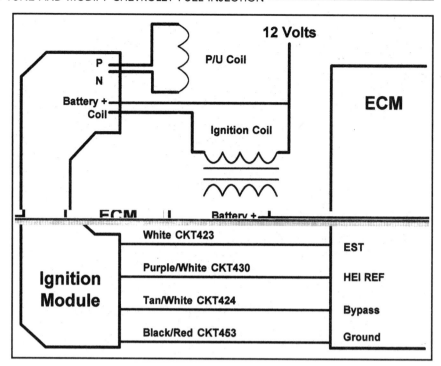

When testing the timing control system there should be 5 volts on the tan and black wire when the engine is running. The black and red wire should be a 0-volt ground. Connecting a dwell meter to the purple and white wire should yield a steady duty cycle; there should be a variable duty cycle on the white wire. This variable duty cycle should increase as the engine rpm increases.

ECM uses to control the switching of the ignition module to control timing. There should be a pulse with a varying duty cycle on this white wire. If the signal is present at the ignition module but there is no timing advance occurring, then the ignition module is defective.

To test for the presence and proper function of this signal, attach a dwell meter (or, if available, a duty cycle meter) to the white wire as close to the ignition module as is practical. As you rev the engine, the dwell shown on the meter should change.

If the dwell changes but the timing does not, check the connections at the ignition module. If the connections are clean and tight, then replace the ignition module.

If there is no change in the dwell reading, and it reads either 0 to 90 degrees Fahrenheit on the four-cylinder scale, then the problem is either in the ECM or the tan and black EST bypass wire.

EST Bypass Wire Testing

Check for 5 volts on the EST bypass wire as close to the ignition module as possible. One convenient place to test would be under the hood at the connector that is disconnected to check initial timing.

If there is 5 volts at this point, then check for 5 volts at the point where the tan and black wire connects to the ignition module. If 5 volts is not present at the ignition module on the tan and black wire but the engine runs, the problem is one of the following:

- An open or ground in the tan and black wire
- A bad connection for the tan and black wire at the ECM
- A grounded (defective) module
- A bad ECM

If there is 5 volts at the ignition module connection but no signal on the white wire, then there is a bad connection of the tan and black wire at the ignition module, or the ignition module is defective.

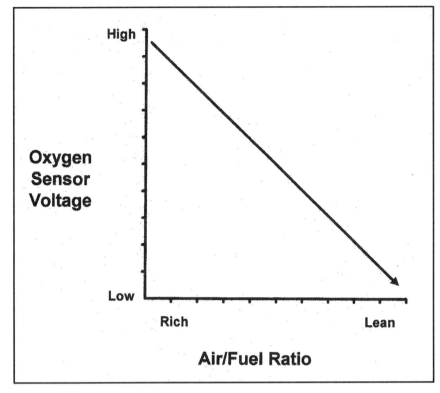

A rich-running engine will produce a high oxygen sensor voltage while a lean-running engine produces a low oxygen sensor voltage.

Minimum air is a measurement, in rpm, of the amount of air entering the engine across the throttle plates with the IAC completely closed.

Air-Fuel Ratio Testing

One of the best things about Chevrolet's late-model fuel injection system is its ability to monitor itself. One of the functions of the fuel injection computer is called the Engine Running Field Service mode. To enter this mode, simply put a jumper between terminals A and B of the ALCL connector with the engine running. When the engine is thoroughly warmed up, the Check Engine light will be flashing on and off about every other second if the air-fuel ratio is correct.

Note: If the Check Engine light is flashing on and off very quickly (about two times per second), then the ECM is operating in the open-loop mode. Allow the engine to warm up further before proceeding with the test procedure.

If the Check Engine light is on more than it is off, then the engine is running rich. Look for problems like these:

- The car needs an oil change. Over a period of time, normal blowby will contaminate the engine oil with unburned fuel. This unburned fuel is then sucked through the PCV system into the intake manifold where it has the net effect of causing the engine to run rich.
- Fuel pressure too high. If the fuel pressure is too high, then every time an injector opens, an excessive amount of fuel is pushed through it and the engine runs rich. Testing of fuel pressure is covered in chapter 7 on troubleshooting.
- There may be a problem in the evaporative canister purge system. If the canister purge valve is damaged or stuck open, then it will permit fuel from the gas tank to enter the intake system.
- Leaking injector. An injector that is not closing properly or that continues to spray fuel even after the ECM assumes it is closed will allow more than the computed amount of fuel to enter the engine.
- Ruptured fuel-pressure regulator diaphragm. A rather odd event

can occur: the diaphragm in the fuel-pressure regulator can rupture, leaking into the vacuum supply hose on the PFI and TPI engines. This can allow considerable amounts of fuel to entire the intake manifold, sometimes without affecting the fuel pressure.

• A defect in the ECM or its sensors. Refer to chapter 4 on how the EFI, PFI, and TPI systems work.

If the Check Engine light is off more than it is on, then the engine is running lean. Look for problems such as these:

• A vacuum leak. Air entering the engine after the throttle plates may be caused by a vacuum leak. In the case of a PFI or TPI, this would be completely unmeasured air. Since it is unmeasured, the ECM does not know to increase the amount of fuel to compensate for the extra air. The end result is that the engine runs lean.

• False air. Any air that enters the intake system undetected by the Mass Air Flow sensor is false air. A vacuum leak would fall into this category, as well as air entering the crankcase through something like a bad gasket on an oil filler cap. False air is not a relevant concern on TBI cars, however. All the air metering on TBI is done after the throttle plates and would therefore be a vacuum leak. For more details on false air see the troubleshooting chapter.

• Low fuel pressure. If the fuel pressure is too low, when the injector opens, very little fuel flows through it and the engine runs lean.

• The oxygen sensor has been tricked! Many things can cause oxygen to enter the exhaust, making the oxygen sensor and therefore the Check Engine light believe that the engine is running lean, when, in fact, it may be running very rich. These factors include an exhaust leak, low compression, a secondary ignition problem, incorrect ignition timing, a defective air pump system, and a defective oxygen sensor.

Check behind the throttle plates on PFI and TPI intake systems for a build-up of carbon and soot. This build-up can result in stalling and hesitation. Clean the throttle bore with a mild solvent on a toothbrush. Be sure to remove the IAC motor and remove any soot.

The Model 220 TBI assembly is a dual-injector throttle body assembly. V-6 and V-8 TBI systems such as those found in light duty truck applications use this throttle body. In order to set up for the minimum air adjustment, follow the same procedure as for the PFI systems. With the key on but the engine off, place a jumper wire between the A and B terminals of the ALCL connector. Wait about 30 seconds for the ECM to move the IAC to its fully extended position; then disconnect the electrical connector, start the engine, and adjust the minimum air.

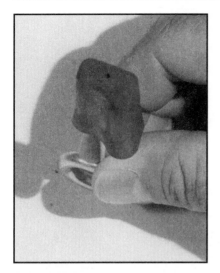

The early single-injector TBI applications, such as the models 300 and 500, use a plug to cut off airflow through the idle air by-pass. After the plug is installed, the minimum air is easy to adjust.

The final step in any minimum air adjustment is to check and reset the closed-throttle TPS voltage. Most Chevrolet applications have a closed-throttle TPS voltage of 0.5 volts. This Tuned Port Camaro has the proper closed-throttle voltage. The TPS adjustment can also be checked where the TPS signal wire enters the ECM or with a scanner.

Oxygen Sensor Testing

The oxygen sensor is used by the ECM to monitor the air-fuel ratio. In spite of what you might hear from different professionals, testing the oxygen sensor is actually quite simple. Connect a high-impedance voltmeter to the oxygen sensor at the point where the purple wire from the ECM connects to the oxygen sensor. Leave the oxygen sensor connected and start the engine. Allow it to run until the ECM goes into the closed-loop mode.

Closed loop can be confirmed by entering the field service mode with the engine running and waiting until the Check Engine light stops flashing two times per second and begins to flash every other second. With the engine idling, watch the voltage reading on the voltmeter. It should be constantly changing back and forth from a voltage between 100 and 400 millivolts to a voltage of 600 to 900 millivolts.

If the oxygen sensor voltage does not change as described, then further testing of the oxygen feedback system is necessary. Refer to the troubleshooting chapter for more details.

Minimum Airflow

Back in the good old days, one of the final touches of any good tune-up was to adjust the curb idle speed. On today's fuel-injected GM cars the idle speed is controlled by the ECM and is not adjustable. However, there is an important adjustment that is similar to curb idle. This adjustment is known as minimum air. The minimum air adjustment determines the amount of air allowed to pass across the throttle plates when they are closed. Incorrectly adjusted minimum air can result in tip-in hesitation, stumbling, and stalling on deceleration.

To begin the minimum airflow adjustment it is necessary to check for throttle body coking. When a PFI engine is shut off, hot crankcase vapors rise through the PCV venting system into the air intake system. As they rise, they carry oil and soot, which rests in the intake system, coating the connector tube between

Far easier to rebuild than most carburetors, the TBI fuel-metering assembly has only three major components and a few O-rings and seals. Unlike a carburetor, assuming clean fuel, the TBI assembly can go for the lifetime of the car without requiring a rebuild.

Fuel-pressure regulator rebuild kits are available. Caution should be used when disassembling the pressure regulator, however. The return valve opening pressure is controlled by a long, relatively heavy spring. When the four cover screws are removed, it comes apart much like the proverbial snake in a can.

The IAC must be disconnected from the ECM during a cylinder balance test. If left connected, the ECM would command it to compensate for the loss of rpm, which indicates relative cylinder power, and the test results would be invalid.

the MAF and the throttle plates. To a lesser extent this coating also builds up on the MAF itself. When the engine is started, incoming air picks up this oil and soot, depositing it at the first low-pressure area it comes to—just behind the throttle plates. Evidence of coking is a ridge of soot built up behind the throttle plates, which can be felt with the tip of the index finger.

Coking behind the throttle plates forms a seal that reduces the amount of air passing across them. Minimum Airflow is affected, causing hesitation and stalling. The Idle Air Control (IAC) passage can also become coked up, causing erratic idle speed control problems.

Cleaning Coking

The best way to clean throttle bore coking is to remove the throttle assembly from the intake manifold. Using solvent and an old toothbrush, scrub the area in front and behind the throttle plates. Remove the Idle Air Control valve from the throttle assembly, dip a rag in solvent and clean the pintle and spring of the IAC. With a small toothbrush, or several pipe cleaners twisted together, scrub out the IAC passage. Reassemble the throttle assembly and re-install on the intake manifold.

Coking can also be a problem on

the sensing element in the Mass Air Flow sensor. The coking forms an insulting blanket over the sensing element which inhibits its ability to accurately measure air flow. A visit to your local electronics shop should yield a spray can of color-television tuner cleaner. Following the directions on the can, spray the sensing element liberally to remove soot build-up.

For maximum results, also scrub the corrugated rubber hose that connects the MAF to the throttle assembly.

It should be noted that throttle body coking is almost never a problem on TBI cars. On TBI there is a constant flow of fuel through the throttle bore which keeps any build-up from occurring behind the throttle plates.

Minimum Airflow Adjustment

Minimum Airflow is the amount of air passing across the throttle plates when the throttle is closed. This specification is given in rpm.

Connect a tachometer to the negative side of the coil. If your engine is distributorless, then you may have to use a special inductive tach or the tachometer on the instrument panel, if the car is equipped with one. You could also connect a high-impedance tachometer to the purple and white wire between the ignition module and the ECM. This

will give you an accurate rpm reading. A scanner can also be used.

The next step is to eliminate the air passing through the Idle Air Control bypass.

PFI and TPI, Models 220 and 700

To adjust airflow on PFI and TPI 220 and 700 models, with key on and engine off, install a jumper between terminals A and B of the ALCL connector. Wait about 30 seconds. During this time the IAC motor will step out to its fully extended position, completely closing off the bypass air. Without turning off the key, electrically disconnect the IAC motor. Then start the engine; it will be running at the minimum air speed. If the minimum air idle speed is not correct, then adjust the throttle stop screw to achieve an idle speed appropriate to your application according to the Minimum Air and TPS Specifications chart in the appendices.

After the minimum air speed is adjusted, connect a voltmeter to the dark blue wire at the Throttle Position Sensor (TPS). If necessary, adjust the TPS to the correct voltage specification by loosening the mounting screws and rotating the TPS until the correct closed-throttle voltage is achieved.

TBI Models 300 and 500

On TBI 300 and 500 models, it is not necessary to trick the IAC into stepping to its fully extended minimum airflow position. A special tool (part number J-33047) is installed into the idle air control passage. This plug ensures that the only air entering the engine is the air passing across the throttle plates. Use the throttle stop screw to adjust for minimum air speed and then the TPS, as necessary.

Crossfire Injection

Crossfire Injection was used on the 1982, 1983, and 1984 Corvette 350 ci (5.7 liter) and Camaro 305 ci (5.0 liter) high-performance applications. Since this system uses two single-unit throttle body injection units mounted on a common intake manifold, adjusting minimum air is a lot like adjusting dual carburetors.

You will need to acquire, or build, a water manometer. This is an extremely sensitive vacuum gauge that reads in inches of water. With the engine warmed up, install the special tool, part number J-33047, in each of the throttle body assemblies; remove the cap from the ported vacuum tube of the rear TBI assembly and attach the water manometer. Adjust the rear assembly throttle stop screw until the manometer reads 6 inches of water (0.45 inHg, or inches of mercury). Now attach the manometer to the front assembly's ported vacuum tube. Repeat the above procedure using the idle balance screw on the TBI interconnect linkage.

Finally, remove the J-33047 tools and adjust the closed-throttle TPS voltage as needed.

Throttle Body Injector Cleaning

The design of the Chevrolet TBI unit lends itself to easy maintenance and cleaning. Kits are available that allow you to disassemble, clean, and overhaul the GM TBI unit much as you would a carburetor.

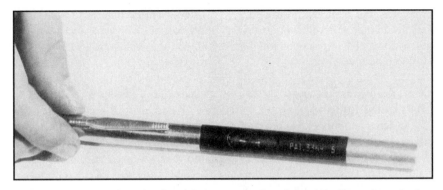

Tools like this one sell for only a few dollars, yet are extremely handy in diagnosing a dead cylinder caused by a secondary ignition problem. Simply hold the tester on a spark plug wire with the engine running. If the neon bulb flashes, then at least some spark is going to the plug. If it does not flash, check for defects in the spark plug, plug wire, distributor cap, and rotor as they relate that cylinder.

Disassembly

First, remove the throttle and cruise control linkages from the throttle bellcranks and the TBI assembly from the intake manifold. Remove the fuel meter body assembly and set aside. It is this unit that will get most of our attention. Remove the fuel inlet nut, the fuel outlet nut, the idle stop screw and spring, the throttle position sensor, and the idle air control motor from the throttle body assembly. What remains of the throttle body assembly can be submersed in carburetor cleaner. Spray a little carburetor cleaner on a shop towel and wipe off the pintle and spring of the IAC motor.

Remove the screws that attach the fuel meter cover to the fuel meter assembly. Gently pry the injector from the fuel meter assembly with two screwdrivers. Use one to pry and the other as a fulcrum. With the injector removed you will see a filter and an O-ring around the end; simply slip these off, spray the end of the injector with carburetor cleaner, and install the new filter and O-ring from the kit. As the injector was removed from the fuel meter assembly, a large, thin, rubber O-ring, as well as a brass ring came off with it. Note that the O-ring installs *after* the brass ring.

Next, invert the fuel meter cover and remove the four screws that hold the fuel-pressure regulator together.

Warning: Both factory and aftermarket service manuals warn not to remove these screws because the pressure regulator assembly includes a large spring under heavy tension which could cause personal injury. The truth is that this spring is only under about 15 lb. of tension; nevertheless, when the last screw comes out of the regulator, the spring will leap out like a snake from a can of peanut brittle.

The final step is to disassemble and replace the diaphragm and reassemble the fuel-pressure regulator.

Reassembly

Remove the throttle body from the carburetor cleaner and reassemble. Be sure to push the IAC pintle in as far as possible before installing it in the throttle body. Also, Minimum Air and TPS must be adjusted before test driving the car.

The ECM might not be able to accurately control the idle speed after the overhaul. If the idle speed is remaining unusually high, it may be necessary to drive the car at a speed greater than 30 mph so that the ECM can reset the IAC.

Port and Tuned Port Injector Cleaning

The cleaning of Port Fuel injectors is done chemically and no disassembly is required. Kits for cleaning Port Fuel Injection systems are available from most national parts and auto mechanic tool distributors. The directions that come with these kits vary by brand, so I will simply suggest reading the instructions for the kit you purchase.

Dead Cylinder Testing: Distributor Ignition

When a late-model fuel-injected engine has a cylinder with a misfire, the effects can go far beyond a mere rough idle or loss of power. A cylinder that is still pulling in air but not burning that air will be pumping unburned oxygen past the oxygen sensor. This confuses the ECM, making it believe that the engine is running lean. The ECM responds by enriching the mixture, and the gas mileage deteriorates dramatically.

There are several effective methods that can be used to isolate a dead cylinder. All of these methods measure the power produced in each cylinder by killing them one at a time with the engine running a little above curb idle.

Back in the good old days, we used to take a test light, ground the alligator clip and pierce through the insulation boot at the distributor cap end of the plug wire. This would ground out the spark for one cylinder and an rpm drop would be noted. The greater the rpm drop, the more power that cylinder was contributing to the operation of the engine. Actually, this is a valid testing procedure; however, piercing the insulation boot is only asking for more problems than you started with.

Another method of performing a cylinder balance was to isolate the dead hole by pulling off one plug wire at a time and noting the rpm drops. The problem with this method is that you run the risk of damaging either yourself or the ignition module with a high-voltage spark.

So, let's explore some valid alternatives.

Cylinder Inhibit Tester

Several tool companies produce a cylinder shorting tach-dwell meter. These devices electronically disable one cylinder at a time while displaying rpm. Engine speed drops can be noted. Unfortunately, these testers can cost $500 or more and most will not do a cylinder balance on a distributorless engine.

Shoestring Technique

Another valid method does the old test light technique one better. Cut a piece of 1/8-inch vacuum hose into four, six or eight sections about 1 inch long each. With the engine shut off, one at a time (so as not to confuse the firing order) remove a plug wire from the distributor cap, insert a segment into the plug wire tower of the cap and set the plug wire back on top of the hose. When you have installed all the segments, start the engine. Touching the vacuum hose conductors with a grounded test light will kill the cylinder so that you can note rpm drop. Again the cylinder with the smallest drop in rpm is the weakest cylinder.

Whichever method you choose for isolating a dead cylinder, follow this procedure for the best results:

- Adjust the engine speed to 1,200–1,400 rpm by blocking the throttle open. Do not attempt to hold the throttle by hand; you will not be steady enough.
- Electrically disconnect the IAC motor to prevent its affecting the idle speed.
- Disconnect the oxygen sensor to prevent it from altering the air-fuel ratio to compensate for the dead cylinder.
- Perform the cylinder kill test. Rpm drop should be fairly equal between cylinders. Any cylinder that has a considerably lower rpm drop than the rest is weak.
- Introduce a little propane into the intake, just enough to provide the highest rpm. Repeat the cylinder kill test. If the rpm drop from the weak cylinder tends to equalize with the rest, then you have a vacuum leak to track down. If there is no significant change in the power output from the weak cylinder, then proceed to the next step.
- Open the throttle until the engine speed is about 1,800–2,000 rpm.

Repeat the cylinder kill test, adding propane. If the rpm drops are now equal, then the most likely problem is that the EGR valve is allowing too much exhaust gas to enter the intake at low speeds. Remove the EGR valve and inspect for excessive carbon build-up and proper seating. If the rpm drop on the cylinder in question remains low, then the problem is most likely mechanical.

- Perform both a wet and dry compression test. If you have low, dry compression and low, wet compression, then the problem is a bad valve or valve seat. If the dry compression is bad but the wet compression is good, then the problem is the piston rings. If the compression is good both dry and wet, then the problem is in the valve-train, such as the camshaft, lifters, or pushrods.
- After testing is completed, reconnect anything that was disconnected or removed for testing.

Dead Cylinder Testing: Distributorless Ignition

If you were to buy a tester capable of doing a cylinder balance on an engine equipped with a distributorless ignition system, you would need to spend thousands of dollars. However, the vacuum hose trick already described will work nicely.

Be sure to disconnect the IAC, the oxygen sensor and stabilize the rpm at 1,200–1,400 for the first balance test. The rest of the testing procedure is exactly what it was for the distributor-type ignition.

Tuning Serial Data

Those who want to dig a little deeper into the thoughts and decisions of the ECM will need to purchase a scanner. The scanner plugs into the ALCL connector under the dash and displays the information that the ECM is receiving from its sensors. It also displays some details about the decisions it is making, and what it is commanding its various actuators to do. A good scanner can cost from a

When performing a cylinder balance test, regardless of the sophistication of the equipment being used, the throttle position should be maintained mechanically. A tapered clothespin half could do this job very well. Or a tool designed to do that job, such as the one shown here, can be purchased for only a few dollars.

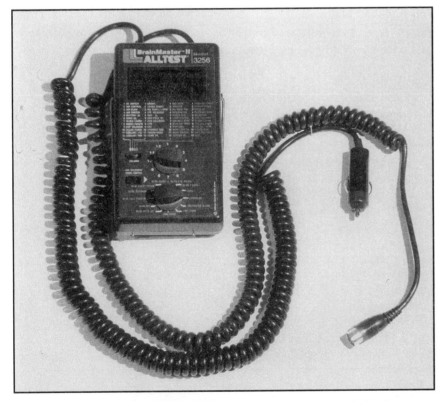

Idle speed, minimum air, TPS voltages, and the ECM's compensation for air-fuel ratio errors can be tested using the diagnostic scanner. At several hundred dollars, this is not a tool for the faint of heart to purchase, and although they are not necessary to troubleshoot most drivability problems, their value cannot be underestimated. Many auto parts stores and virtually all automotive tool dealers either stock or can order diagnostic scanners.

few hundred to several thousand dollars; thus it is not a purchase to be taken lightly but should be considered by the serious mechanic. Through the scanner, it is possible to diagnose driveability problems and determine what corrective actions the ECM has made in its fuel injection or timing control program.

Integrator

The Integrator is one of the key pieces of serial data information. First of all, let us refer to an imaginary dictionary where the Chevrolet engineers have translated phrases into numbers. If we looked up the phrase "normal for the Integrator," we would find it translated into the numbers "128±6." With the engine warmed up and in closed loop, we would request that the scanner display the Integrator or INT function. Numbers from 122 to 134 would be a normal reading. If the number displayed is greater than 134, then it indicates the fuel injection system is responding to a request for more fuel. The oxygen sensor must be sensing a high oxygen content in the exhaust, which it interprets as a lean condition. The ECM assumes the engine is running lean and enriches the mixture. Keep in mind that a lean air-fuel ratio is only one of the possible causes of a high oxygen content in the exhaust.

If the Integrator number is less than 122, it indicates that the engine is receiving extra fuel from somewhere. The Integrator is therefore compensating by decreasing the amount of fuel that the injector(s) is delivering to the engine.

The Integrator changes in direct response to changes in exhaust oxygen content. Whenever the engine is in open-loop operation, the Integrator number will be fixed at 128. Once the engine enters closed-loop operation, the Integrator will be constantly changing to correct minor errors in air-fuel ratio.

If the Integrator number is high, look for a source of extra oxygen in the exhaust such as a vacuum leak, restricted injector(s), cracked exhaust manifold, defective air pump upstream-downstream switching valve, mechanical engine problems, low fuel pressure, or defective ignition plugs, cap, rotor, or plug wires.

If the Integrator number is low, look for a source of extra fuel such as a defective evaporative canister purge system, leaking injector, contaminated crankcase, high fuel pressure, or leaking cold-start injector (where applicable).

Block Learn Multiplier

The Block Learn Multiplier, or BLM, is a long-term version of the Integrator that is stored in memory. The compensating that it does is for longer term trends in air-fuel ratio. Because the Block Learn is stored in memory, it even has the ability to compensate for air-fuel ratio problems while the engine is in open loop, providing the problem was detected the last time that the engine ran in closed loop.

Just like the Integrator normal is 128 ± 6, a higher number indicates that the ECM is adding more fuel to compensate for a perceived lean condition. A lower number indicates the ECM is compensating for a perceived rich condition. Any abnormal reading in the BLM indicates that the perceived need to adjust the air-fuel ratio has been there for a minimum of several seconds. Several seconds may not seem like a long-term situation, but do not forget we are dealing with a computer.

Always keep in mind when troubleshooting that rich and lean on any modern oxygen-feedback fuel injection system is determined by the oxygen content of the exhaust gases, not by fuel entering the engine.

Oxygen Sensor Volts

As displayed on the scanner, oxygen sensor voltage should be constantly switching from a low of 100–300 millivolts to a high of 600–900 millivolts voltage and from a high voltage to a low. Should the oxygen sensor voltage be mostly on the high side, then the engine is running a little rich. Do not become concerned if the oxygen sensor voltage and the Integrator/Block Learn disagree at times. These three functions respond at different rates.

Crosscounts

Crosscount readings indicate the number of times that the oxygen sensor voltage has switched or crossed from a lean reading to a rich or from a rich reading to a lean. A minimum of four between scan samples on a regular basis is good. The number could be as high as 30 or 40 crosscounts.

Oxygen Sensor Status

Oxygen sensor status indicates whether the general trend of the oxygen sensor is to read rich or lean. It should change back and forth slowly, maybe every 10 to 15 seconds.

TPS Volts

During a scan test for tune-up purposes, we are only concerned with the closed-throttle voltage that was discussed in the minimum air section of this chapter. Refer to the Minimum Air and TPS Specifications chart in the appendices for the proper values.

Gps Airflow

Gps (grams per second) airflow relates only to Port Fuel and Tuned Port Injection applications. This is a measurement of the amount of air sensed entering the engine though the MAF. At an idle, this should be about 7 gps. If the reading is low, then it could be an indication of false air entering the engine.

False air can be found by shutting the engine off and stuffing a rag over the end of the MAF. Blow compressed air at about 5–10 psi into the intake manifold. Using a spray bottle filled with soapy water, spray around the rubber connector boot between the MAF and the throttle assembly. Wherever bubbles rise up is false air. False air can also come in the form of a vacuum leak (after the throttle plates, by definition), so spray the entire intake system with the soapy water. Make repairs as needed.

Coolant

Engine coolant temperature can be read on the ECM in either Celsius or Fahrenheit or both, depending on the scanner.

Engine Speed

The ECM registers crankshaft rpm (revolutions per minute).

IAC

The IAC (Idle Air Control) data gives the position of the IAC motor. There are 256 positions that this motor can be in. The lowest number is 0, the highest is 255. The lower the number is, the slower the ECM tries to make the engine run. Seven to 20 are typical at idle for most cars. If the number is at or close to 0, this could indicate a vacuum leak or incorrectly adjusted Minimum Air.

Tuning Serial Data

Serial data	Normal	Problem
INT	128	155
BLM	128	128
O$_2$	100–900mV	100–400mV
Crosscounts	4–16	1
O$_2$ status	Rich	Lean
TPS volts	0.50	0.50
Gps airflow	7	3
Rpm	800	775
Coolant	203°F	203°F (95°C)
IAC	15	0

For the problem readings, first look at the Integrator data: 155 indicates that the ECM is adding fuel to compensate for what the oxygen sensor believes is a lean exhaust condition. The normal Block Learn Multiplier number indicates that this is either a relatively new problem or an intermittent problem that has occurred very recently, perhaps within the past minute.

The oxygen sensor voltage does not seem ever to be indicating rich, a fact that is further confirmed by the low number of crosscounts. The TPS reading is normal, but the gps airflow is low. Lean exhaust, high Integrator, and low gps point toward false air or a vacuum leak.

```
                              DIACOM

 ┌─────────────────────────────────────┬─────────────────────────────────────┐
 │ VEHICLE TYPE: 1989 CHEVROLET        │ ENGINE TYPE: 5.7 Liter V8           │
 ├─────────────────────────────────────┼─────────────────────────────────────┤
 │ ECM MODE STATUS: NOT LINKED         │ ECM PROM ID: 0                      │
 └─────────────────────────────────────┴─────────────────────────────────────┘

   Engine Speed................    0 RPM   Integrator..................    0 #
   Desired Idle Speed..........    0 RPM   Block Learn Multiplier......    0 #
   Vehicle Speed...............    0 MPH   Block Learn Cell Number.....    0 #
   Coolant Temperature.........  -40 F     Battery Voltage.............  0.0 VDC
   Start-up Coolant Temp.......  -40 F     Fuel Pump Relay Voltage.....  0.0 VDC
   Manifold Air Temperature....  -40 F     Engine Running Time.........    0 Min
   Throttle Sensor Voltage..... 0.00 VDC   Cooling Fan Duty Cycle......    0 %
   Filtered Load Variable......    0 g/c   Canister Purge Duty Cycle...    0 %
   Oxygen Sensor Voltage.......    0 mV    EGR Duty Cycle..............    0 %
   Idle Air Position...........    0 #     EGR Diagnostic Switch.......  Off
   Air Flow Rate...............    0 GPS   Closed Loop Status..........  Off     m
   Injector Pulse Width........  0.0 mS    Fuel Mixture Status.........  Lean    o
   Spark Advance...............  0.0 DEG   Park/Neutral Switch.........  Off     r
   Knock Retard................  0.0 DEG   Power Steering Switch.......  Off     e
   Spark Control Counts........    0 #     3rd Gear / OD Switch........  Off     %

  Configure   Disk   Exit   Info   Link    Mode   Options   Print    Trbl_codes
 Establish communication with vehicle
 COMMAND=> USE   ARROW KEYS or SPACE BAR TO SELECT OPTION THEN PRESS "Enter"
```

The serial data stream provides several pieces of tune-up related information. Among these are Integrator, Block Learn, oxygen sensor voltage, oxygen sensor crosscounts, oxygen sensor status, TPS volts, gps airflow, coolant temperature, engine rpm, and Idle Air Control valve position.

Serial Data Example

Serial data	Problem
INT	128
BLM	85
O_2	100–900 mV
Crosscounts	8
O_2 status	Rich
TPS volts	0.50
Gps airflow	8
Rpm	800
Coolant	203°F (95°C)
IAC	13

The most noteworthy thing about this example is the BLM. It indicates that the engine has been running rich for a while, yet the BLM has fully compensated for the condition and no longer requires the assistance of the INT. The fact that the O_2 voltages are crossing the full range further confirms this. The O_2 status is currently indicating rich, but at any moment could swing lean. Grams per second airflow is good, as well as TPS voltage.

For more in-depth information on serial data interpretation, refer to the chapter on troubleshooting.

Tuning Procedure

To keep your fuel-injected engine well-tuned and running smoothly, follow these steps, whether you do it yourself or have it done by a professional.

- Replace spark plugs
- Replace air filter
- Replace fuel filter
- Inspect distributor cap
- Inspect distributor rotor
- Test plug wires
- Check or set initial timing
- Check for timing advance
- Clean the MAF sensor
- Clean throttle body coking
- Check/adjust minimum air
- Check air-fuel ratio control

The 100,000-Mile Tune-Up

Regulations and mandates will require manufacturers to assure the Environmental Protection Agency (EPA) that an engine is capable of pushing a car down the road for 100,000 miles without violating emission standards before a tune-up is required. My own experience indicates that, with the state of technology we currently see, this is a very realistic goal. When I first became an automotive mechanic in the early 1970s, this would have been a ridiculous goal. I figured I was doing really well when a tune-up would last 15,000 to 25,000 miles. There are several things that have contributed to this improved tune-up parts' life expectancies.

Electronic ignition systems deliver a much more powerful punch to ignite the fuel. This means that even deteriorated secondary ignition components can fire the air/fuel charge effectively. Electronic fuel injection does a much better job distributing the air/fuel charge than the carburetor. Combustion temperatures are, therefore, more even and combustion quality is improved. Fuel-fouled spark plugs are virtually unheard of today. Similarly, the high temperatures resulting from lean combustion almost never damage spark plugs.

C
O
N
T
E
N
T
S

**SYMPTOM:
ENGINE CRANKS
BUT WON'T START 76**

**SYMPTOM:
ENGINE DIES OR
STUMBLES ON TIP-IN
ACCELERATION 80**

**SYMPTOM:
ENGINE CUTS AND
MISSES AT IDLE OR
LOW SPEEDS 83**

**SYMPTOM:
DELAYED OR
EXTENDED START 85**

**SYMPTOM:
LACK OF POWER 86**

**SYMPTOM:
HUNTING IDLE 87**

**SYMPTOM:
SURGE AT
CRUISE SPEEDS 88**

**SYMPTOM:
DIESELING OR
RUNNING-ON 88**

**SYMPTOM:
EXHAUST ODOR 88**

**SYMPTOMS:
DETONATION,
KNOCKING OR
PINGING 89**

**SYMPTOM:
POOR FUEL
ECONOMY 89**

CHAPTER SEVEN

FUEL INJECTION TROUBLESHOOTING

Troubleshooting Chevrolet fuel injection systems falls into three broad categories: Diagnosis by symptom; use of trouble codes; and use of the diagnostic scanner.

Trouble codes are delivered upon the request of the technician whenever a problem is either occurring or has occurred within the last 50 restarts of the engine. These codes indicate that the computer has detected a complete failure in one of the monitored circuits. Not all of the electrical and electronic circuits attached to the ECM are monitored for codes, and most driveability symptoms will be caused by things that will not set a code. For this reason we will begin by looking at driveability symptoms and their causes. An effort has been made to list the causes of the symptoms and testing procedures in the order of expense and likelihood.

Symptom: Engine Cranks but Won't Start

Five things are necessary in order to get an internal-combustion gasoline engine to start:
- Sufficient engine cranking speed (which is necessary to create compression)
- Compression
- Air
- Spark
- Fuel

The following is a logical procedure to use in testing for a no-start condition.

Engine Cranking Speed Testing

Probably one of the quickest and easiest (although not very scientific) ways of checking engine cranking speed is by sound. This is especially effective if you are working on your own car or a model with which you are familiar. The use of a tachometer for this test is good since it will not only tell you what the cranking speed of the engine is, but will also tell you the primary ignition is not working if you do not get a reading.

Traditionally, the tachometer is connected to the negative side of the coil. If you cannot find the negative side or if you are working on a distributorless engine, connect your digital tach to the purple and white wire that runs from the ignition module to the ECM.

Compression Testing

Compression is a function of engine cranking speed and proper cylinder sealing. Like checking engine cranking speed, the skilled technician can often tell by the sound of the engine whether or not there is sufficient compression to start. If there is any doubt as to the quality of compression, a proper compression test should be done.

Low compression on one or two cylinders usually is not enough to keep the engine from starting—a no-start condition would require that several of the cylinders have low compression. Low compression in all cylinders might be caused by a jumped timing chain or stripped timing gears. If the compression is good

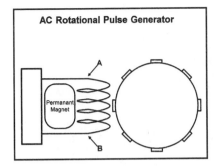

AC Rotational Pulse Generator

The pickup coil used in the EST distributor system is an AC pulse generator. Although many experts recommend testing this device with an ohmmeter, this is actually an incomplete test. A better test is to use an AC voltmeter. With the engine cranking, the output voltage of the pickup coil should be at least 0.5 volts.

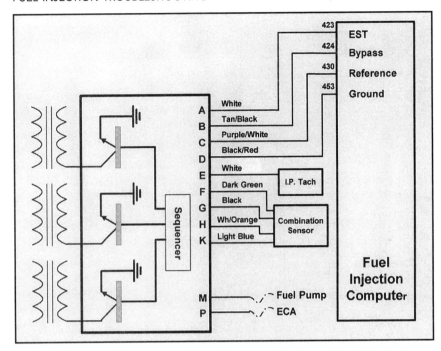

General Motors introduced distributorless ignition through the Buick, Oldsmobile, Cadillac Division in 1984. A Hall Effects sensor called the combination sensor is located on the crankshaft pulley and provides information about engine rpm and crankshaft position. There are two 12-volt power sources and the same white, tan and black, purple and white, and black and red wires and signals as found in the distributor EST system.

in some cylinders but not in others, the most likely cause is a blown head gasket, worn rings, or bad valves.

Airflow Testing

To a large extent, the ability of the engine to bring air into the combustion chamber is dependent on the same things that give the engine compression, as well as its intake system design, the condition of the exhaust system, the condition of the air filter, the Idle Air Control valve, and the throttle plates.

Testing this system involves a visual inspection. Check the condition of the air filter. If it appears dirty or restricted, replace it. On PFI and TPI it is especially important to check the air tube that connects the MAF to the throttle assembly for evidence of false air. Inspect that the IAC valve and throttle plates are not coked. Also inspect the throttle for free movement.

On the opposite side of the combustion chamber from the intake system is the exhaust. In order for the intake to be able to pull air in, the exhaust must be capable of getting air out. Have someone crank the engine while you hold your hand over the tailpipe. There should be noticeable pressure against your hand. If you are unsure whether or not there is

enough, compare it to the amount being pushed out the exhaust while cranking a car that will start.

Spark Testing: Distributor Ignition

In order to test for spark, the first step is to insert a screwdriver into the spark end of a plug wire and hold the screwdriver about 1/4 inch from a good ground. Crank the engine and check for spark.

If there is a spark, remove one of the spark plugs, insert it into the plug wire, place the plug on a good ground, such as the intake manifold, and crank the engine. If the spark plug sparks and does not appear worn, proceed to the Fuel section of this chapter. If there is no spark at the spark plug, replace the plugs.

If there was no spark from the spark plug wire, the second step is to probe the negative terminal of the ignition coil with a test light. If you are not sure which side of the coil is negative, then insert it first on one side, crank the engine, then insert it on the other side. One side of the

coil should have a steady supply of switched ignition voltage while the other side should flash on and off as the engine is cranked. If neither has power, then you have an open wire in the voltage supply to the coil. If both sides have steady power, then the problem is the primary ignition's control of the coil.

In the old days, the first thing to look at would have been the points; today, we have to look at the equivalent of the points: the pickup coil, the ignition module, and the interconnecting wiring.

The next step is to test the pickup coil. The variable reluctance transducer is the standard type that Chevrolet has used since the debut of electronic ignition in 1975. This pickup produces an AC signal. Disconnect the connector at the pickup and connect an AC voltmeter. Be sure that nothing will get tangled or damaged when the engine is cranked. Crank the engine and watch the voltmeter. If it measures a voltage of at least 0.5 volts AC, then the pickup is good.

The C31 distributorless ignition system uses a Hall Effects sensor such as this to determine crankshaft speed and position. Connect a digital tachometer or dwell meter to test for the presence of an output signal. The tach should read something other than zero while the engine is being cranked. The dwell meter should read something other than zero or full scale.

This is the underside of the C31 ignition module showing its mounting studs. The wire hanging below the module is the 12-volt power connection for two of the coils.

The next step is to check the wiring and test the ignition module. Other than replacing with a known good unit, there is no accurate way to test an ignition module without special equipment. There is a trick of the trade, however, that will confirm that a module is probably good, but it will not determine if it is bad. Remove the distributor cap, or access the pickup coil terminals (P and N) of the ignition module whatever way is necessary. Lick the tip of your index finger and touch the P and N terminals simultaneously. Use caution because if the ignition module is good, the ignition coil will discharge up to 50,000 volts. Although the current from this discharge is so low that it cannot seriously injure a healthy person, the voltage is high enough to startle you into injuring yourself.

With all of the steps completed, if the pickup produces 0.5 AC minimum, the problem is in the wiring or module. If the module triggers through the wet-touch test, it is not the module. But if it does not trigger, the module might still be good. If it triggers the coil, carefully inspect the wiring harness; if it does not trigger, inspect the wiring harness and replace the module if no problem is found in the harness.

Spark Testing: Computer Controlled Coil Ignition

The Computer Controlled Coil Ignition, or C31 ignition system, was developed for the Buick, Oldsmobile and Cadillac division of General Motors. This ignition system may be found on Chevrolet V-6 applications. Like the distributor applications, if there is no high-voltage spark, the fault could lie in the coil, ignition module, or engine position sensors.

The C31 system uses Hall Effects sensors to monitor the positions of the crankshaft and camshaft. To test the crankshaft, probe the black and light green wire at the crankshaft position sensor connector. This wire runs from terminal H of the ignition module to terminal A of the crankshaft position sensor and provides power for the sensor. The reading should be 10 volts or more with the ignition on.

Probe terminal B on the crankshaft position sensor connector with an ohmmeter. Although there will be some resistance measured, it should not indicate infinite resistance. If it does, there is an opening in the wiring harness or the module is defective. Measure the resistance in the black and yellow wire that runs from terminal B to terminal G of the

ignition module. The resistance should be very close to zero. If the resistance is close to infinity, then repair the wire. If the wire is good, check the connection at the ignition module. If the connection is good, replace the module.

Next, check the voltage with the key on at terminal F of the ignition module, a blue and white wire. The voltmeter should read either close to 0 volts or over 5 volts. When you rotate the crankshaft, the voltage reading should change. If it was 5, it should drop to 0; if it was 0, it should rise to 5.

To monitor the position of the camshaft, with the key on there should be 10+ volts in the yellow wire running from terminal N of the ignition module to terminal C of the cam sensor. Test for this at the cam sensor.

The ground for the sensor is a black and pink wire running from terminal B of the sensor to terminal L of the ignition module. Measuring with an ohmmeter from the B connection to ground should read resistance but not infinity. Like with the crank sensor, an infinity reading means that you need to check the wiring and connections. If they check out to be good, then replace the ignition module.

A brown and white wire runs from the A terminal of the cam sensor

The early C31 coil packs consist of three coils all in one block and are replaced as a unit.

to the K terminal of the ignition module. This wire carries camshaft position information to the module in order to sequence the injectors. The voltage should read either 5+ or 0. Rotate the crankshaft. If the voltage was low, it should go high; if the voltage was high, it should go low. If it does not change, replace the crank sensor.

If the voltage at terminal F of the ignition module switches high and low as the crankshaft is rotated, the crank sensor is good. If the voltage at terminal K of the ignition module switches high and low as the crankshaft is rotated, the cam sensor is good.

There should be 12 volts on the pink and black wire connected to the P and M terminals of the ignition module. There should be continuity to ground through ignition module terminals G and L.

If all of this hold true, then replace the ignition module.

Note: If the application you are working on is equipped with a four-wire combination sensor on the crankshaft and no sensor on the camshaft, there will be a blue wire to terminal K and a dark green wire to terminal F. The diagnostic procedure is the same. Also, terminals J, L, and N will be unused.

Spark Testing:
Direct Ignition System

Chevrolet's answer to the C31 system is the Direct Ignition System, or DIS. DIS is used on four-cylinder and six-cylinder applications. Compared to the C31 system, DIS is very simple. A purple and a yellow wire carry an AC signal from the crankshaft sensor to the ignition module. On the ignition module there are three terminals where these wires connect. The purple wire connects to terminal A and should have 0.7+ volts AC on it when the engine is being cranked. The yellow wire attaches to terminal C and should also have 0.7+ volts AC on it when the engine is cranked. Terminal B of this connector is shielded and should have no signal on it.

There is also a two-wire connector on the ignition module. A pink and black wire is connected to terminal B and should have 12 volts on it when the ignition switch is turned on. There is also a black and white wire attached to terminal A; this is the ground.

If there is 12 volts with the key on at terminal B of the ignition module and a ground at terminal A of the two-pin connector, then check for an AC voltage of at least 0.7 at both A

and C of the three-pin connector. Then replace the ignition module.

These diagnoses for no spark on C31 and DIS assume that none of the coils are producing a spark. If even one of the coils produces a spark, then the problem is most likely a defective coil or coils.

Quick Check Tips
- If there is no spark but the injector pulses, then the lack of spark is a result of a secondary ignition problem such as a bad coil, distributor cap, or rotor. If there is no spark and the injector does not pulse, then the problem is likely a primary ignition problem.
- A quick and easy test for pulses from C31 cam and crank sensors is to connect a dwell meter to terminals F and K of the ignition module. If the sensors are good, the dwell meter will read somewhere between 10 and 80 degrees Fahrenheit on the four-cylinder scale. If not, it will read 90 or 0 degrees Fahrenheit.

Fuel Pump Testing

When the key is turned to the run position but the engine is not cranked, the fuel pump will run for about 2 seconds and then shut off. If you can hear the fuel pump run for these few seconds, then you know that the fuel pump relay and the ECM's control of the relay are operative.

Connect a fuel-pressure gauge to the Schraeder valve on the fuel rail of the PFI and TPI applications. No such valve is provided on the TBI applications, so it will be necessary to T the gauge into the inbound fuel line. Crank the engine for several seconds; the fuel-pressure gauge should indicate more than 30 but less than 45 psi for Port Fuel and Tuned Port Injection, and between 9–15 psi for Throttle Body. This pressure should hold from several minutes to hours slowly bleeding off. If the pressure drops off quickly, then the fuel-pump check valve, fuel-pressure regulator, or an injector is leaking.

If the pressure is low, run a volume test with and without the fuel filter.

Disconnect the fuel filter on the outbound side. Using an approved fuel container, install a hose on the outbound end of the fuel filter and crank the engine for about 15 seconds. The pump should flow a minimum of a pint. If the flow is less, remove the filter and repeat the test. Install a new filter if the flow is good when the filter is removed but reduced when it is reinstalled. If the pressure is low but the flow is adequate, replace the fuel-pressure regulator.

If the fuel pressure is correct, check to see if the injectors are opening. There are several good ways to do this, but the best way is to use a stethoscope to listen for them to open and close while the engine is being cranked. If a mechanic's style stethoscope is not available, a piece of 1/2-inch heater hose held near the ear and next to the injector will work almost as well.

If the injectors are not clicking and there is no spark, repair the ignition system first. If there is spark but the injectors do not click, check for voltage while cranking at the pink and black wire that supplies voltage for the injectors. If voltage is present, connect a tachometer to the negative side of the injectors. Cranking the engine should show some sort of a reading (it doesn't matter what the reading is as long as it is not 0). If there is no reading, connect a dwell meter set on the four-cylinder scale to the HEI (high energy ignition) REF input to the ECM and crank the engine. The dwell meter should read about 45 degrees Fahrenheit. If the reading remains high or low (90 or 0 degrees Fahrenheit), then check continuity in the purple and white wire from the ignition module to the ECM.

On engines equipped with C31 it is also necessary for the ECM to receive a signal from the Cam sensor. Connect a dwell meter to the black wire that runs from the ignition module to the ECM. Although the reading on this wire will be very erratic, we can consider it okay if the reading is anything other than 0 or 90.

If the pink and black wire is car-

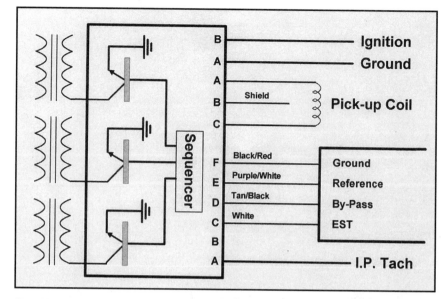

Chevrolet's version of distributorless ignition is called DIS. An AC pickup is used to sense crankshaft position. There is power and ground to the ignition module and the same four timing control wires found in the distributor systems.

rying 12 volts to the injectors, if there is a pulse on the HEI REF wire, if the wires are good, and on the C31 system there is a pulse on the black cam sync wire, check for continuity from the injectors to the appropriate ECM terminal (such as D15 or D16 for the 1987 5.7-liter TPI Corvette).

Symptom: Engine Dies or Stumbles on Tip-In Acceleration

Back in the days of carburetors, points and pony cars, this symptom would have pointed toward a bad accelerator pump, an interruption in primary ignition as the breaker plate moved from application of vacuum advance, a low float level, or secondary ignition problems. The same general areas should be addressed on GM fuel-injected cars. For this symptom the items we are going to check are:

• Spark plugs and plug wires
• Distributor cap and rotor (where applicable)
• Fuel pressure at idle and on acceleration
• MAP or MAF signal (the modern equivalent of vacuum advance)
• TPS signal (the modern equivalent of the accelerator pump)

Since spark plugs and plug wires, distributor cap and rotor are covered in chapter 6 on tune-ups, refer to the appropriate sections for test procedures.

Fuel Pressure Testing

Attach the fuel-pressure gauge on the PFI and TPI applications to the Schraeder valve; T the gauge into the inbound line to the throttle body on TBI. Start the engine, allow it to run for a few seconds, and then check the fuel pressure. PFI and TPI engines should show 30+ psi; TBI and CFI engines should register 9+ psi.

If the fuel pressure is low, rev the engine while watching the fuel-pressure gauge. The PFI and TPI systems should have an immediate increase of about 5–10 psi in fuel pressure when the throttle is cracked open. Although there should be no increase in fuel pressure on TBI and CFI, there should be no loss of pressure either. If the fuel pressure performs properly under these conditions, it does not rule out the possibility that a loss of fuel pressure is at fault.

Running the engine and snapping the throttle does not create nearly the fuel demands that actual acceleration

This is the module from Chevrolet Division's own distributorless ignition system, the DIS or Direct Ignition System. This module is from a 2.8-liter V-6 application. The electrical connections sticking up from the module are where the coils mount. Found mounted on the front side of the block, servicing of the module/coil assembly is easiest from under the car.

DIS coils can be changed individually. Note that each coil has two secondary output towers. Spark is delivered simultaneously to two cylinders, in this case 2 and 3. It fires the cylinder that is on compression in much the same way as a distributor system, while the plug that is on the exhaust stroke fires at a much lower voltage. Coils may be stamped with cylinder numbers as the one above; however, no problems will occur if the 2 and 3 coils are moved to a different position on the module as long as the plug wires are not moved with them.

creates. Secure the fuel-pressure gauge in a safe place where it can be seen while driving the car. Take the car out for a test drive and watch for a noticeable decrease in fuel pressure as the hesitation or stalling occurs. If the problem is stalling, be sure to notice if the decrease in pressure occurs before or after the engine dies. It is normal for the fuel pressure to drop after the engine dies.

If the fuel pressure fails to increase on the PFI and TPI or if the pressure drops, the problem is insufficient supply volume. Check or replace the fuel filter. Also inspect the lines for kinking damage or restriction.

MAP Testing

The MAP (Manifold Absolute Pressure) sensor measures the load on the engine. As the engine load increases, the voltage output from the MAP sensor also increases. This increase in voltage should follow the increase in load in a direct fashion. If for some reason the change in MAP voltage should hesitate or falter, then the engine will stumble or die.

Connect an analog voltmeter to the center wire of the MAP where it connects to the ECM. Position the voltmeter so it can be safely viewed while driving or, ideally, take along an observer. The voltmeter reading should increase as you accelerate; the voltage should change as the engine load changes. Have someone watch the voltmeter carefully as the car stumbles or dies. Does the voltage increase all the way to 5 volts or drop all the way to 0? If it does, inspect the wires thoroughly.

With the voltmeter still connected, tug gently on the wires to the MAP sensor, then wiggle the wires and connectors. If the voltage does not peg high or low when you wiggle the wires but did peg high or low as you drove the car, then replace the MAP sensor. If the voltage did fluctuate as you wiggled the wires, repair the damaged wire.

MAF Testing

The MAF (Mass Air Flow) sensor is one of the primary sensors in the calculation of the Load Variable. The Load Variable is calculated by the ECM on applications that do not use a MAP sensor. Like the MAP sensor, a fault in the MAF or MAF circuit can cause a stumble or hesitation.

With the engine warmed up and idling, gently tap on the MAF. If the engine rpm drops noticeably or if the engine dies, replace the MAF.

Bosch MAF Testing

The 5.0- and 5.7-liter TPI cars have Bosch MAF sensors. If the tap test is passed, connect an analog voltmeter where the signal wire of the MAF connects to the ECM. This is usually a dark green wire connected to ECM terminal B12. As you rev the engine the voltage on this wire should gradually increase as the airflow increases. Keep in mind that due to variations in volumetric efficiency, the volume of air flowing into the engine does not vary exactly with changes in throttle position and therefore the voltage will not change at the same rate you change the throttle opening.

As on cars equipped with a MAP sensor, connect the voltmeter to the dark green wire and position the meter where it can be safely viewed

The MAP sensor on this 1983 2.0-liter Citation is located in the center of the firewall. Although this is a typical location, you might have to hunt around a bit to locate it in the engine compartment of another application. The MAP sensor measures engine load, producing a voltage that ranges between 0.8 volts during deceleration to over 4.5 volts in the key-on, engine-off mode. Except for when the engine is off, the higher the MAP output voltage, the greater the engine load.

The Bosch-style MAF is used on Tuned Port Injection engines and sends a variable voltage signal to the ECM. Beginning at approximately 1 volt at an idle, the voltage should steadily increase as the engine rpm and airflow increase.

by a passenger. As you drive the car, the MAF voltage should vary along with engine vibration and load. Have the observer watch carefully for the voltmeter to peg either high (5 volts) or low (0 volts) as the engine stumbles or dies. If it does, then either the MAF or MAF wiring harness is defective.

If you have access to a diagnostic scan tool, at an idle the MAF should be sensing about 5 to 8 gps (grams per second) of air. Gps airflow should change using the same procedure described for voltage.

Delco MAF Testing

The Delco MAF delivers a square wave to the ECM. Its frequency varies as the mass of air entering the engine varies. With key on and engine off, the frequency is about 32 Hz, and at an airflow well beyond what the typical engine can actually produce the frequency tops out at 150 Hz. The tap test applies to the Delco-style MAF sensor as well

as the Bosch. It must be stressed that this test involves a light tap, not a heavy rap.

Connect a digital tachometer set on the four-cylinder scale to the yellow wire attached to terminal B of the MAF. At an idle the tachometer should read 1,200 to 1,600 rpm (40 to 52 Hz). As you open the throttle to increase the speed of the engine, the frequency should increase as well. Although it will not necessarily follow throttle position precisely, the general trend of frequency should be about the same as the position of the throttle.

A test drive with an observer to watch the tachometer should find the MAF frequency tracking close to power demand and load on the engine. Should the frequency drop to 0 or leap to 150 Hz (4,500 rpm on the four-cylinder scale) as the engine hesitates, stumbles, or dies on acceleration, check the MAF wiring. If the MAF wiring is okay, then replace the MAF.

Throttle Position Sensor Testing

The Throttle Position Sensor (TPS) takes the place of the carburetor's accelerator pump. The TPS is instrumental in providing the initial burst of fuel necessary to provide a smooth transition from a stop to power.

Checking the TPS is much easier than testing the MAF. Connect a digital voltmeter to the dark blue wire attached to the TPS and turn the ignition switch to the run position. Refer to the Minimum Air and TPS Specifications chart in the appendices for the correct closed-throttle voltage. This closed-throttle voltage is critical to smooth acceleration on fuel-injected Chevrolet products. If the closed-throttle voltage is correct, open the throttle to wide open. The voltage should gradually increase to over 4 volts.

If you have any doubts about the smoothness of the voltage increase, connect an analog voltmeter to the TPS signal wire (dark blue) and slowly open the throttle (key on, engine off) while observing the voltmeter. A smooth transition from the closed-throttle voltage to the wide-open-throttle voltage is essential for smooth acceleration.

Miscellaneous

Other things that can cause hesitation or stalling on tip-in acceleration are:

- Water contamination in the fuel. Disconnect the outbound line of the fuel filter and prepare to capture a small quantity of fuel in a clear plastic container. Cycle the ignition switch on and off several times to energize the fuel pump. Allow the captured fuel to sit in the container undisturbed for about 30 minutes. If there is a high concentration of water in the gasoline, it will settle to the bottom of the container. The corrective action is to drain and refill the tank using a water-purging additive that can be obtained at your local auto parts store.

- False air (especially between the MAF and throttle assembly, where applicable) requires a careful visual inspection of the corrugated rubber tube that connects the MAF to the throttle assembly.

- Alternator output voltage lower than nine volts or greater than 16 volts. Connect a voltmeter to the output terminal of the alternator, start the engine and check the voltage at both idle and part throttle.

- A bad ground on the black and red wire of the ignition module. Visually inspect this wire for both tightness and lack of corrosion.

- EGR opening prematurely or leaking. Disconnect the vacuum hose from the EGR valve and plug it. Test drive the car. If the symptom is gone, inspect the vacuum hose for proper routing. If the routing is correct, replace the EGR solenoid. If the problem persists, remove the EGR valve and inspect to see that it is sealing properly when no vacuum is applied.

- Canister purge valve opening prematurely or leaking. With a pair of locking pliers, clamp off the canister purge hose and test drive the car. If the symptom is relieved, replace the canister purge solenoid.

Symptom: Engine Cuts and Misses at Idle or Low Speeds

When an engine cuts out and misses at idle or low engine speeds, one or two cylinders are not supplying an amount of power equal to the rest. Isolating the offending cylinder can be done as described in the tune-up chapter, or by using a pair of sissy pliers (non-conductive pliers designed for removing spark plug wires with the engine running) to locate the weak cylinder. The cylinder(s) that does not produce a significant rpm drop when the wire is removed is the offending party.

After locating the bad cylinder, the following items must be checked:

- Plug wires
- Spark plugs
- Compression
- Fuel pressure
- Injector wiring
- Injector balance
- Water contamination in the fuel
- Valve action and timing

Plug Wire Testing

There are basically two tests to be performed on the spark plug wires: The "rainy day" test and the resistance test.

To perform the rainy day test, fill a spray bottle with water, start the engine and allow it to warm up. Spray the plug wires with a fine water mist. When you spray a wire with leaking secondary insulation, the misfire will get worse.

For the second plug wire test, shut off the engine, then remove, test, and reinstall one plug wire at a time. Using an ohmmeter set on the 100,000 ohm or greater scale, check the resistance of each wire you remove. If any are above 30,000 ohms, replace them.

Spark Plug Testing

In a very practical sense, without a good ignition oscilloscope the best way to test spark plugs is to swap the plug on the weak cylinder with a plug from a good cylinder. If the bad cylinder moves, then the spark plug was at fault and should be replaced.

Compression Testing

Refer to chapter 6 on tuning for a thorough discussion of compression testing.

Fuel Pressure Testing

Test the fuel pressure as described earlier in this chapter. Refer to the fuel-pressure chart for the correct pressure for your car. These pressures should increase 5–10 psi when the engine is revved.

Fuel Pressures

PFI	33–38 psi
TPI 5.7 liter	35–39 psi
TPI 5.0 liter	42–46 psi
CFI	9–13 psi
TBI	9–13 psi

Injector Wiring Testing

Testing the injector wiring harness electronically can be difficult on some applications. A practical alternative is to use a mechanic's stethoscope, touching each of the injectors and listening for the distinctive click of their opening and closing. Should any of the injectors fail to click, exchange that one with one of the good ones. If the nonclicking injector moves, you have a bad injector. If it does not move, then inspect and repair the bad wiring harness.

Injector Flow Testing: TBI and CFI

Throttle Body Injection should be checked for a good spray pattern. A poor spray pattern can cause a weak cylinder in the same way a bad injector wiring harness can. Since the injector spray pattern on TBI can be seen while the engine is running, connect a timing light to the engine and hold it so that it shines on the injector spray. Since the timing light and the injector(s) are synchronized to primary ignition, the spray will appear to stop in mid-air for close inspection. The spray should be cone shaped and aimed straight downward from the center of the injector.

Connecting a voltmeter to the output of the Delco MAF should yield a steady, unchanging voltage of about 2.5. Anything else and there is a problem with the MAF or MAF wiring harness. This easy test does not confirm accuracy, however. Connect a digital tack on the four-cylinder scale to the output wire. At an idle, the tach should read about 900 to 1,200 rpm (30 to 40 Hz) and steadily increase as the engine rpm increases.

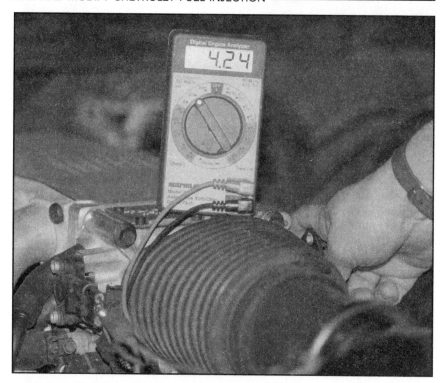

As the throttle is opened, the TPS signal voltage should smoothly increase to over 5 volts. Here, wide-open throttle voltage is being tested on a 5.0-liter TPI Camaro.

Injector Flow Testing: PFI and TPI

Port Fuel and Tuned Port Injection are a little tougher to deal with on the issue of injector flow testing. Back in the old Bosch D-Jetronic days we used to raise the injectors above the intake, place each one of them in a graduated cylinder, then crank the engine for 30 seconds while watching for a cone-shaped pattern. After cranking, we would inspect the fuel levels in each of the graduated cylinders to determine if an equal amount of fuel was passing through each injector. PFI and TPI engine design does not facilitate this type of flow testing.

There is an alternate method of performing the injector flow test that requires a special tool known as an Electronic Fuel Injector (EFI) tester. This device costs between $100 and $300 and can be obtained through any of the tool truck dealers. The tester pulses the injector between 1 and 500 times when activated.

First, connect a fuel-pressure gauge to the PFI or TPI fuel rail. Then connect the battery leads of the EFI tester to the battery and the Number One injector. Cycle the key once or twice in order to pressurize the fuel system to between 16 and 19 psi. If the pressure is too high, bleed it down to the proper range before testing the injector. Set the tester for 100 pulses of 5 milliseconds and press the button; the pressure indicated on the fuel-pressure gauge will drop. Record the amount of pressure loss.

Repeat this procedure with each injector, recording the pressure drops. Re-test any injector that has a large variation from the others. Any injector that shows more than 1.5 psi difference in pressure drop, either higher or lower than the rest, is either restricted, leaking, or not closing properly. Clean the injectors with one of the injector cleaning systems that are available. These systems start at about $100 and go up. Or simply replace the offending injector(s).

Water Contamination Testing

Use the procedure described earlier in this chapter for avoiding tip-in acceleration stalling.

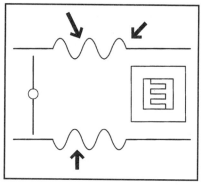

The PFI and TPI systems that use a MAF use a connecting boot between the MAF and the throttle assembly. A hole or loose connection on the boot can allow unmeasured air to enter the engine. When air enters the engine without being measured, the ECM does not put in enough fuel and the engine tends to run lean. Up to a point, this can be compensated for by the inputs from the oxygen sensor.

Valve Action Testing

The compression test checked for valves that did not seal properly. If you have not found the cause of the miss by

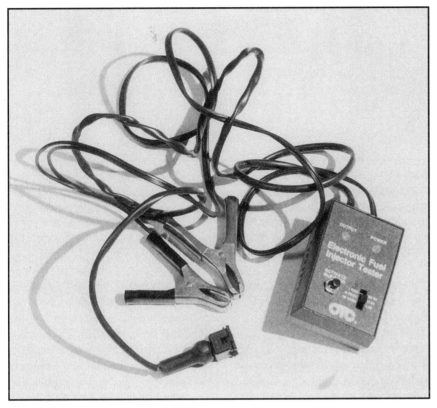

The fuel-injector tested is used to test the relative flow rates through the injector on PFI and TPI applications. The alligator clips are connected to the battery, the injector harness to the injector. After pressurizing the fuel system without the engine running to around 35 psi, touch the Activate button and the injector will be pulsed between 1 and 500 times, depending on the position of the switch on the right on the tester. Record the pressure drops; they should be even on all cylinders. The injector that has a smaller pressure drop than the rest is restricted.

now, then it is time to remove the valve cover(s) and measure the pushrod lift. If the movement of the pushrods varies from one cylinder to another, then the camshaft and lifters need to be replaced. Note: Exhaust and intake valves may have different lifts from each other, but all of the intakes and all of the exhausts should be the same.

Symptom: Delayed or Extended Start

When the symptom is delayed or extended start, the engine cranks okay and it eventually does start but the cranking time is several seconds rather than an immediate start-up. Possible causes are listed as follows:
- Driver's starting procedure
- TPS sticking or binding, sending a high voltage to the ECM
- High resistance in the coolant sensor circuit
- Incorrect fuel pressure

- EGR stuck open
- Deteriorated secondary ignition
- Fuel in the fuel rail draining back into the tank, either through the pump check valve or through the system pressure regulator
- Leaking injectors
- Inoperative cold-start injector (5.0-, 5.7- and some 2.8-liter engines)

Starting Procedure

TBI, CFI, PFI, and TPI are all designed to start the engine with the foot off the accelerator. Pressing the accelerator just a little while cranking the engine greatly increases the amount of fuel passing through the injectors, and could flood the engine. Pressing the accelerator even farther (80 percent or more) will put the ECM into its Clear Flood mode. Clear Flood virtually shuts down the injectors and will make an engine that is not flooded difficult to start.

The correct starting procedure is to crank the engine with your foot off the gas pedal until the engine starts.

TPS Testing

Check the TPS for sticking or binding when sending the high voltage to the ECM. Check the TPS output voltage (the dark blue wire). If the closed-throttle is above 0.85 volts, either adjust or replace the TPS.

You should note that some non-adjustable (there are no adjustment slots) Throttle Position Sensors may be as high as 1.2 volts and still be acceptable.

Coolant Sensor Testing

Check the coolant sensor for high resistance. Backprobe the yellow wire at the coolant temperature sensor; the voltage should be less than 4.8 but greater than 0.1 volts. If the starting problem is primarily a cold-start problem, then use an ohmmeter to measure the resistance of the sensor with the engine cold. If the starting problem is primarily during hot starts, then measure the coolant sensor resistance with the engine near operating temperature and use the chart shown.

Coolant Sensor Resistance
Engine Temperature Resistance

(degrees F)	(ohms)
210	185
160	450
100	1,800
70	3,400
40	7,500
20	3,500
0	25,000
-40	100,700

Fuel Pressure Testing

Test the fuel pressure as described in the section on engine cutouts and misses in this chapter.

EGR Testing

Remove and visually inspect the EGR valve to ensure that the pintle is seating properly and that it is not stuck open.

The most common cause of a loss of power in a fuel-injected car is a dirty air filter.

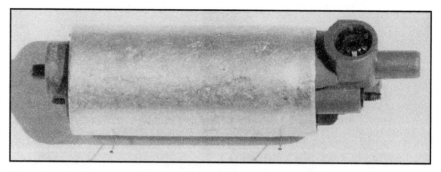

The fuel pump has a check valve to prevent fuel draining from the fuel rail or throttle body when the engine is shut off. Connect a fuel-pressure gauge to the system and watch how quickly the fuel pressure drops. If the pressure reaches zero within a couple of minutes of engine shutdown, pinch off the return line from the fuel-pressure regulator to the tank and repressurize the system by cycling the ignition key. If pressure continues to drop, replace the fuel pump; if not, replace the pressure regulator.

Secondary Ignition Testing

Inspect the cap, rotor, spark plugs, plug wires, and coil wire as described in the tune-up chapter and earlier in this chapter.

Fuel Rail Drainage Testing

If fuel is not held in the fuel rail in-between operation of the vehicle, then it is necessary for the pump to fill and pressurize the fuel system before the car will start and run.

There are two components through which the fuel can drain back into the tank. The first is the fuel-pump check valve. Connect a fuel-pressure gauge, start the engine and let it run for a few minutes. Shut off the engine and monitor the residual pressure for 30 minutes or so. During this time the residual pressure should drop very slowly and not reach zero. It is normal and acceptable for the pressure to drop to zero in a matter of hours, however.

If the pressure drops, repressurize the fuel system and clamp off the outbound line from the tank. If the fuel pressure remains high, replace the fuel pump because it has a bad check valve.

The second component through which fuel can drain back, is the system's pressure regulator. If the pressure drops when the outbound line from the fuel tank is clamped off, repressurize the system and clamp off the return line to the tank. If the pressure does not drop, replace the pressure regulator.

If the pressure continues to drop, you may have leaking injectors or a bad cold-start injector.

Injector Leakage Testing

If clamping off the outbound and return lines from the gas tank did not stop the loss of pressure, the only place the fuel could be leaking is through the injectors and into the intake manifold. Remove the fuel rail assembly from the intake manifold and reconnect it to the fuel lines so that the injectors are visible and the fuel rail can still be pressurized by the fuel pump. Pressurize the fuel rail and observe the tips of the injectors. Any injector that forms a drop of fuel heavy enough to fall off and then begins to form another one can be considered leaking.

If a leaking injector is found (including the cold-start injector on applications having one), using an injector-cleaning system may help, or you might simply replace the bad injector(s).

Cold-Start Injector Testing

The cold-start injector (on 5.0-, 5.7- and some 2.8-liter engines) will only affect the starting of the engine when both the engine temperature and the ambient temperature are below 95 degrees Fahrenheit. Use a mechanics stethoscope on cars equipped with a cold-start injector to determine if it is working. A single click should be heard and assuming that the system has fuel pressure, the cold-start injector is working.

If no click is heard, check for voltage on the tan wire at the thermo-time switch. Check first while cranking the engine with its temperature below 95 degrees Fahrenheit. The voltage should be very low or at 0 for 5 to 12 seconds, then go high. If it does not, then replace the thermo-time switch. Start and warm the engine, shut off the engine and again check the voltage on the tan wire while cranking the engine. The voltage should be high (about 12 volts). If it's low, replace the thermo-time switch.

The voltage supply for this circuit is through the crank fuse, which should also be checked.

Symptom: Lack of Power

If the car does not have the expected or usual amount of power, check for the following:
• Dirty air filter
• Restricted fuel filter
• Contaminated fuel
• Low fuel pressure
• ECM grounds
• EGR failing to close
• Restricted exhaust
• ncorrectly adjusted throttle linkage
• Alternator voltage lower than 9 or higher than 16 volts
• Low compression
• Incorrect valve timing
• Worn camshaft
• Problems in the secondary ignition

All of the possible causes of the symptoms listed have been discussed except for dirty air filter, ECM grounds, restricted exhaust, and incorrectly adjusted throttle linkage.

Air Filter Testing

A dirty air filter is probably the easiest of all diagnostic tasks. Remove the air filter from the air cleaner and inspect it. If the air filter is dirty, replace it. If in doubt, replace it.

ECM Testing

The ECM is grounded to the engine block; a bad connection at the ground can cause any number of different driveability symptoms, including a lack of power. Inspect all ground connections on the engine block. On some applications these grounds are well hidden and pose a real challenge to finding and inspecting them. Also check the ground cable from the battery to the engine.

Exhaust Testing

A restricted exhaust can reduce the engine's ability to breathe just as much as a dirty air filter. The only effective way to test for excessive exhaust backpressure is with a backpressure gauge. Although this gauge can be purchased through automotive tool dealers, it can also be made by taking your oxygen sensor down to the local hardware store and purchasing whatever pipe fittings are necessary to adapt a carburetor-style fuel-pressure gauge to the hole where the oxygen sensor installs. (Note: When reinstalling the oxygen sensor, coat the threads with a good anti-seize compound such as GM part number 5613695.)

Install the exhaust backpressure gauge into the oxygen sensor hole. Start the engine and allow it to idle. With the engine at normal operating temperature, the gauge reading should not exceed 1.25 psi.

Now raise and hold the engine rpm over 2,000 and observe the gauge. If the reading is in excess of three psi, the exhaust has excessive backpressure.

Often, even among knowledgeable, professional mechanics, the most common component to condemn for excessive backpressure is the catalytic converter. While this is a likely suspect, it is by no means the only component to be at fault. Also inspect the mufflers and exhaust pipes. Remember, many of the exhaust components, including the catalytic converter, carry a five-year/50,000 mile warranty (and even more in some jurisdictions).

Throttle Linkage Testing

Remove whatever is necessary to view the throttle plates in the throttle assembly. Have someone fully depress the throttle while inspecting to see that the throttle plates are fully open. If they are not, adjust the throttle linkage as necessary. Also look for any obstruction to the full movement of the throttle, such as the carpet or floor mat being bunched up under the accelerator pedal.

Symptom: Hunting Idle

Hunting idle is a symptom distinguished from a miss in that the idle speed varies in slow surges over a wide rpm range. Possible causes of hunting idle include:
- Bad motor mounts
- Sticking or binding throttle linkage
- TPS sticking or binding
- Incorrect or fluctuating alternator output voltage
- Misadjusted or defective park/neutral switch circuit (also known as neutral safety switch)
- Defective or incorrect PCV valve
- Vacuum leak
- Defective or erratic evaporative control system
- Power-steering pressure switch input erratic
- ECM ground
- EGR valve not seating properly
- Inconsistent fuel pressure
- Uneven compression
- Problems with the air conditioner (if the problem only occurs with the A/C on)
- Erratic Idle Air Control System

As for the previous symptoms, only the diagnostic methods not already discussed will be covered here.

Motor Mount Testing

If the motor mounts are broken, this will allow the engine to move as it runs. This movement could put stress on the throttle cable or linkage, pulling the throttle open and closed. Checking the motor mounts is done by prying on them with a large pry bar. If you are able to unseat the engine off of its mounts as you do this, the mount is bad and should be replaced.

Park/Neutral Switch Testing

Backprobe the orange and black wire at the neutral safety switch (which is usually located near the base of the steering column) where the wire goes through the firewall. In park or neutral there should be 0 volts on this wire. In all other gears there should be 12 volts.

If the voltage fluctuates as the engine surges, adjust the park-neutral switch. If adjustment does not eliminate the voltage fluctuation, replace the switch.

If there is 0 volts both in park or neutral and in other gears, disconnect the park-neutral connector and test for 12 volts on the orange and black wire. If there is 12 volts, replace the park-neutral switch.

Should the voltage on the orange and black wire be 0 when the park-neutral connector is disconnected, then trace the wire back to the ECM. There should be 12 volts on the orange and black wire at the ECM with the park-neutral connector disconnected. Typical ECM terminals for the orange and black wire (depending on the application) are D11, D16, BD16, B10, and W18. If there is 12 volts at the ECM, repair the open circuit in the orange and black wire.

If there is not 12 volts at the ECM, inspect the wire for a grounded condition. If the orange and black wire is not grounded, then the ECM is defective.

Note: Before replacing the ECM, always double- and triple-check your test results. Also, your local dealer may be able to replace the ECM at little or no charge under emissions warranty. Check with the service manager before purchasing a new ECM.

PCV Valve Testing

With the engine idling, remove the PCV valve from its normal position in the valve cover or intake manifold. Leave the PCV connected to its vacuum hose. Place your thumb over the end of the PCV; if the surging idle ceases, replace the PCV valve.

Power-Steering Switch Input Testing

Backprobe the light blue and orange wire at the power-steering pressure switch with a voltmeter. There should be 12 volts with the engine idling and no load on the power-steering system. If the voltage fluctuates between 12 and 0 volts as the idle surges, check the power-steering fluid level. If the level is light, replace the power-steering pressure switch. If the new switch behaves the same way, replace the power-steering pump.

Turn the steering wheel to full lock either direction. The voltage should drop to 0 and remain there without fluctuating as long as the steering is held at full lock.

If the light blue and orange wire has 0 volts on it with no load on the steering, remove the connector from the power-steering pressure switch. If the voltage goes to 12 volts, replace the switch. If the new switch behaves the same way, replace the power-steering pump.

If the light blue and orange wire has 0 volts on it when removed from the power-steering pressure switch, trace the wire back to terminal C8, W6, BD12, D16, or D11, depending on the application. At the ECM if there is 12 volts, repair the open circuit in the light blue and orange wire.

If there is 0 volts at the ECM with the power-steering pressure switch harness disconnected, then inspect the light blue and orange wire for a ground. If the wire is not grounded, then replace the ECM.

Idle Air Control Testing

Refer to chapter 8 for information concerning Code 35.

Symptom: Surge at Cruise Speeds

The symptom of surging at cruise speeds involves variations in engine power and load even when the throttle is held in a steady position. Possible causes include the following:

- Abnormal operation of the automatic transmission converter clutch or the A/C compressor clutch
- Vacuum leak
- Restricted fuel filter or lines
- Erratic fuel pressure
- Secondary ignition

All of these topics were covered earlier in the chapter.

Symptom: Dieseling or Running-On

The condition of dieseling or running-on occurs when the engine continues to run or chug even after the ignition switch has been shut off. There is only one thing that can cause this to happen: Fuel is getting into the intake system from somewhere. Sources of this fuel are:
- Leaking injectors
- Leaking cold-start injector (5.0-, 5.7- and some 2.8-liter PFI only)
- Defective canister purge system
- Crankcase severely contaminated with fuel
- Fuel-pressure regulator leaking into the vacuum hose

The first two problems were covered earlier in this chapter.

Canister Purge System Testing

If the canister purge valve is allowing fuel vapors to leak into the intake manifold as the engine is shut off, it will supply the fuel necessary to cause the engine to diesel. Crimp the canister purge hose (the largest one running from the canister purge valve to the intake manifold). Then run and shut off the engine several times. If the dieseling is gone, replace the canister purge valve.

Crankcase Contamination Testing

If the crankcase is severely contaminated with fuel vapors, these vapors can get into the intake as the engine is shut off and cause the engine to diesel.

Remove the PCV valve from the valve cover or intake manifold, run and shut off the engine several times. If the dieseling is gone, change the oil.

Fuel-Pressure Regulator Leakage Testing

On rare occasions the diaphragm in the fuel-pressure regulator will get a slight tear, allowing fuel to be sucked into the intake manifold. This of course applies only to the systems that raise fuel pressure as the intake manifold pressure increases (PFI and TPI).

The required repair is the replacement of the fuel-pressure regulator.

Symptom: Exhaust Odor

Due to a wide range in the quality of fuel available today, in some cases rotten-egg exhaust odor is unavoidable. This is because all fossil fuels contain sulfur. The quantity of sulfur in each gallon of gasoline varies depending on where it originated and where it was refined.

The first thing that should be done to get rid of a rotten-egg smell is to try several different brands of gas. If the problem persists regardless of brand, then the problem is likely to be that the engine is running rich. If that's the case, check the following:
- Fuel pressure
- Canister purge system
- Leaking injectors
- Leaking cold-start injector (5.0-, 5.7- and some 2.8-liter PFI only)
- Vacuum leak(s)
- Ruptured diaphragm in the fuel-pressure regulator
- Refer to chapter 8 for details on codes 44 and 45

Backfire

A backfire occurs when fuel ignites in either the intake or exhaust. Proper diagnosis requires a feel for whether the backfire occurred in the

A hunting idle can be caused by a defective IAC (arrow), or the ECM and IAC attempting to compensate for problems such as broken motor mounts, an erratically operating charging system, a misadjusted park-neutral switch, a vacuum leak or many other engine and engine peripheral systems.

intake or the exhaust. Check the following for possible causes: secondary ignition, vacuum leaks, exhaust leaks, valve timing or air pump system.

The first four problems were covered earlier in this chapter.

Air Pump Testing

Some of the cars that use an air pump system will be programmed to pump air into the exhaust manifold or catalytic converter under normal cruise conditions, but will divert the air to the atmosphere on deceleration. Clamp off the hose that runs from the diverter valve to the pipe running down to the exhaust. Road test the car. If the backfire is gone, replace the diverter valve.

Symptoms: Detonation, Knocking or Pinging

If you were to take a couple of marbles, drop them in a coffee can and give it a good shake, you would come close to simulating the sound of detonation. Generally speaking, this condition is caused by excessive heat in the combustion chamber. Some possible causes include:

- Overheating engine
- EGR valve not opening
- Detonation sensor system inoperative
- Poor fuel quality

Overheating Engine

Although the engine temperature gauge or warning light is designed to alert the driver if the engine is overheating, this system monitors only one point in the cooling system. There can be places in the engine that are hotter than the engine temperature warning sensor. A visual check for the following should eliminate the possibility of overheating being the cause of detonation: low coolant level, restricted airflow through the radiator, defective radiator electric cooling fan, or defective or loose belt on belt-driven cooling fan.

EGR Valve Testing

The function of the EGR valve is to cool the combustion process to reduce the emissions of oxides of nitrogen. A by-product of this function is to reduce combustion temperatures so that detonation will not occur. If the EGR valve fails to open,

the excess combustion temperatures may cause detonation.

Begin testing this system by ensuring that there is 12 volts with key on and engine off at both the pink and black wire, and that the gray wire is connected to the EGR control solenoid.

If there is 12 volts on both wires, start the engine and ground the gray wire. When you open the throttle, the EGR valve should also now open. On some applications, as soon as the gray wire is grounded the engine will die; either way the EGR valve is working.

If the EGR valve fails to open, check for vacuum to and from the EGR control solenoid. If there is no vacuum into the control solenoid with the throttle cracked open, check for correct vacuum hose routing as per the underhood decal. If there is vacuum in but no vacuum out (with the gray wire grounded), replace the EGR control solenoid.

If there is vacuum to the EGR control solenoid and to the EGR, but the EGR valve does not open, replace the EGR valve.

Detonation Sensor System Testing

If the detonation sensor circuit is inoperative a Code 43 will be set. Follow the Code 43 diagnostic routine in chapter 9.

Fuel Quality

Detonation can also result from poor fuel quality today just as in the days of the big iron land rockets of the 1960s. This symptom might be eliminated simply by changing fuels.

Symptom: Poor Fuel Economy

If your car is averaging poor fuel economy, there are several possibilities to consider:
- Condition of tune-up
- Air conditioner compressor not cycling off
- Vacuum leaks
- Poor engine condition
- High fuel pressure
- Restricted exhaust
- Driving habits

CONTENTS

ALCL CONNECTOR 90

TROUBLE CODES 91

SERIAL DATA 98

SUMMARY 107

CHAPTER EIGHT

ONBOARD DIAGNOSTIC TROUBLE CODES

Chevrolet has designed the ECM to monitor many of its essential circuits and warn the driver by means of a Check Engine light when a failure occurs. This onboard diagnostic system stores a two-digit code in the ECMs memory relating to the circuit where a problem has been detected. The code can be obtained with no special equipment through the Check Engine light. This light is activated by means of a diagnostic connector known as an ALCL or ALDL, supplied by Chevrolet. The ALCL is located inside the car usually above the driver's left knee.

ALCL Connector

The ALCL connector is a 12-way connector used by the technician to communicate with the ECM. Through the ALCL the technician can command the ECM to enter any of four diagnostic modes. He or she can also bypass certain actuating functions such as fuel pump relay, air pump switching valve, and the Check Engine light. In addition, information can be delivered to the technician through the ALCL by two different means: trouble codes and serial data. Serial data is computer-coded information that can only be interpreted with a scanner.

An important thing to note as we look at the purpose of each of the ALCL terminals is that not all applications will have actual connections in each of the 12 terminal pockets. The only two that always have connectors in them are terminals A and B.

Terminals A and B

Terminal A is a ground. Terminal B, on the other hand, is the actual self-diagnostic terminal. Grounding terminal B (usually to terminal A) will put the ECM into the field service mode. The field service mode has two forms: key on with the engine off and with the engine running. During the engine-off field service mode, trouble codes are repeatedly flashed out by the ECM through the Check Engine light. With the engine running in the field service mode the ECM will display oxygen sensor status through the Check Engine light.

If the light is flashing on and off very fast, about two times per second, that indicates that the ECM is operating in the open-loop mode. After running for several minutes, the flashing of the light will slow down considerably. At this point the Check Engine light will be displaying oxygen sensor status. The light on indicates that the oxygen sensor is detecting a rich exhaust condition; light off indicates that the sensor is detecting a lean exhaust condition.

Placing a 3,900 ohm (3.9 kilohm) resistor between terminal B and ground terminal A will put the ECM into the back-up mode. This mode activates the limp-home system and the ECM will be using only rpm, throttle position, and coolant temperature to control the injectors. This mode is automatically entered when a major component failure is detected by the ECM. Placing the 3.9 k resistor across terminals A and B tests the ECM's ability to enter this mode.

Placing a 10,000 ohm resistor across terminals A and B will put the ECM into the special mode. This mode will force the ECM into closed loop, ignoring the run timers, and fixes the idle speed at about 1,000 rpm with a fixed IAC position. The only practical purpose it would serve in troubleshooting the typical driveability problem is if the problem was an unstable idle. Entering the special mode would fix the IAC position if the idle speed continued to fluctuate. Then you would know for sure that the problem was not erratic IAC motor control.

The final diagnostic mode is entered by connecting nothing to terminal B. This is the open or road test mode. It is also the mode in which the car is normally driven.

Terminal C

Terminal C is found on some fuel-injected Chevrolets that use an air pump. This terminal is connected to the wire that the ECM uses to ground and activate the air switching valve to allow air to be pumped upstream of the oxygen sensor. During the warm-up process the sensor is preheated by the air pump pumping air across it. Tremendous heat is generated when the carbon monoxide- and hydrocarbon-laden exhaust gases react with the air from the air pump. This is used to hasten the oxygen to a usable temperature. Therefore during warm-up the voltage at terminal C should be low, and once at operating temperature the voltage should be high.

A further use for this terminal is to connect a voltmeter to the oxygen sensor while monitoring the oxygen voltage, ground terminal C. If the voltage drops as terminal C is grounded, it proves that the oxygen sensor is responsive and that the air switching (upstream/downstream) valve and solenoid work.

Terminal D

Quite often, the D terminal is missing, but when it is there it is connected to the Check Engine light.

Applying 12 volts to terminal D will test the Check Engine light for a burned-out bulb. Also, connecting terminal D to any actuator circuit will allow the Check Engine light to be used as a test light.

Terminal E

Terminal E may very well be the most important terminal in the ALCL connector. This is where serial data is transmitted to the scanner. Through this terminal it is possible to get an insight into what the computer is thinking and how it is making decisions. Later in this chapter we will take an extensive look at the use of serial data information in troubleshooting.

Terminal F

The status of the torque converter clutch can be monitored at terminal F. Anytime the brake pedal is not depressed and the torque converter has not been requested to engage by the ECM, there will be 12 volts at terminal F. The voltage will drop to 0 when the torque converter clutch is engaged by the ECM, or the brake pedal is depressed. Depressing the brake pedal removes the power supply from the lock-up solenoid. Once the brake pedal is released the voltage at F will go to 12 volts, then drop to 0 again when the converter clutch is re-engaged.

Terminal G

On some fuel-injected cars where there is a connector in terminal G, it is possible to energize the fuel pump for testing purposes by applying 12 volts. Other applications have this connector in the engine compartment.

Terminals H, J, K and L

Terminals H, J K and L are not used for the fuel injection system.

Terminal M

Some applications with high-speed computers use terminal M in addition to E for the transmission of serial data.

The communication link between the technician and the ECM is the ALCL connector, which is usually located above the driver's right knee. Placing a jumper such as a paperclip between the upper right two terminals A and B with the engine off but the key on will cause the Check Engine (or Service Engine Soon) light to flash out trouble codes. With the engine running, an illuminated Check Engine light indicates a rich exhaust, while the light off indicates a lean exhaust. Placing a 3.9 kilohm resistor between terminals A and B puts the ECM into fuel back-up mode; a 10 kilohm puts it into special mode.

Trouble Codes

Chevrolet has three types of codes that can be extracted through the Check Engine light. First there are hard codes. These are the only codes that will turn on the Check Engine light with the engine running in the road test mode. If the Check Engine light is on while driving, there is at least one hard code stored in the diagnostic RAM.

There are also soft codes and an indicator code. The codes are obtained by entering the key on/engine off field service mode by placing a jumper wire between terminals A and B of the ALCL. The codes will be flashed out through the Check Engine light three times each beginning with a Code 12. Code 12 is the only indicator code. This identifies that the ECM recognizes the fact that the engine is not running and that the ECM has entered the field service mode.

After all the codes have been flashed out three times each, the

ECM will again flash out Code 12 three times, then continue to repeat the codes until the field service mode is exited by removing the jumper.

After retrieving the trouble codes you must clear the codes by removing the fuse that powers the diagnostic RAM. This fuse can be found in one of three places. The first is in the fuse panel labeled ECM B or ECM 11. The second is in a fuse holder located in the engine compartment. It can be recognized by the large orange wire going into at least one side of the fuse. The final location is a fusible link coming off the positive terminal of the battery.

After the codes are cleared, reconnect the fuse and start the engine. Run the engine for 5 to 10 minutes until the Check Engine light comes back on. Retrieve the codes again; as before, the only codes you will get this time will be the hard codes. These codes are the ones you want to diagnose first. Begin by troubleshooting the lowest-numbered code first, unless there is a 50s code, such as 51 or 55.

Soft codes indicate intermittent problems that are not currently detected. These are the codes that did not return after running the engine and pulling the codes a second time.

Trouble Code Explanations

Following is a list of trouble codes and explanations. The number in the "[]" indicates the simplified troubleshooting chart to follow. The trouble code charts may be used instead.

Trouble Code Definitions

Code	Trouble Indication
12	No tach signal to ECM
13	Oxygen sensor
14	Coolant sensor circuit voltage low [S3]
15	Coolant sensor circuit voltage high [S3]
21	Throttle Position Sensor voltage high [S5]
22	Throttle Position Sensor voltage low [S5]
23	Manifold air temperature voltage high [S3]
24	Vehicle Speed Sensor
25	Manifold Air Temperature voltage low [S3]
32	EGR system malfunction
33	Mass Air Flow sensor reading too high [S6] or MAP too high [S5]
34	Mass Air Flow sensor reading too low [S6] or MAP too low [S5]
35	Idle Speed Control error
36	MAF burn-off problem (Bosch MAF only)
41	Interruption in tach signal to ECM [S7] or [S8] depending on application
42	Electronic timing control (EST) problem
43	Knock sensor (ESC) system problem
44	Lean exhaust condition
45	Rich exhaust condition
51	Defective PROM or PROM installation
52	CALPAK missing
53	High battery voltage
54	Low-fuel pump voltage
55	Defective ECM
63	MAP too high (2.8L Generation II only)
64	MAP too low (2.8L Generation II only)

Code 12

Technically, Code 12 means that the ECM is currently receiving no reference pulses from the ignition system. Since the trouble codes are pulled with the engine not running, the ECM is not receiving distributor reference pulses. This code is not stored in memory and its practical use is to identify that the ECM is in the self-diagnostic mode. This is particularly important when there are no faults and when Code 12 is the only code that will be flashed out.

Code 13

Code 13 relates to the oxygen sensor circuit. Typically, this code will set when:
- The engine has been running at least 2 minutes
- The oxygen sensor signal voltage is steady between 0.33 and 0.55 volts

- The TPS signal is above 6 percent (this equates to about 1,200 rpm)

Code 13 is most often set when there is an open circuit in the oxygen sensor circuit. A sooted sensor can also set the code.

Code 14

Code 14 denotes a shorted Coolant Temperature Sensor (CTS). This code will set anytime the CTS indicates a temperature above 266–284 degrees Fahrenheit. The high temperature indication means that the voltage has dropped to nearly 0 for at least 2 to 4 seconds (depending on the application). The only things that could cause this are a grounded yellow wire to the CTS or a shorted sensor.

Problems in the coolant temperature sensor circuit can affect the operation of these functions:
- Fuel delivery, air-fuel ratio
- Electronic spark timing
- Radiator cooling fan
- Torque converter clutch
- Idle air control

Resistance in the CTS decreases as the temperature goes up. When troubleshooting the sensor, use the following resistance specifications. These specs should be used as a guideline. Do not expect your test results at a given temperature to precisely match the readings given in the chart.

Coolant Sensor Resistance

Engine temperature (degrees F)	(degrees C)	Resistance (ohms)
210	99	185
160	71	450
100	38	1,800
70	21	3,400
40	4	7,500
20	-7	13,500
0	-18	25,000
-40	-40	100,700

Code 15

Code 15 denotes an open coolant temperature sensor circuit. When the computer sees more than about 4.8 volts on the yellow wire to the coolant temperature sensor, a Code 15 will be set providing the engine has been running for a minute or so. This voltage can occur as a result of an open circuit anywhere in the coolant sensor circuit. As in Code 14, this fault will affect the accuracy of fuel delivery, timing, radiator fan, torque converter clutch, and the Idle Air Control system.

Code 16

Code 16 denotes system voltage out of range—if the control of the charging system does not limit the voltage to less than 16 volts, that is. Applications that use this code will disable all the actuator solenoids as the code is set, to protect the ECM from the excess voltage surges that will be created when the solenoid is turned off and the magnetic field in them collapses. The cure for this code lies in the repair of the charging system.

Code 21

Code 21 denotes Throttle Position Sensor (TPS) voltage high. This code occurs if the ECM sees a signal above 2.5 volts from the TPS while the MAP sensor is sending a signal indicating a pressure less than 8.25 psi in the intake manifold. This pressure would equate to a manifold vacuum of greater than 16.75 inches. Additionally, the engine rpm may need to be less than 1,600. These three conditions must be met for 2 to 5 seconds.

Typically, a Code 21 will be set when there is a defective TPS, when the black ground wire running from the TPS to the ECM has an open circuit or when the dark blue TPS signal wire is shorted to voltage. Check for these three problems.

Code 22

TPS voltage low. Code 22 occurs when the ECM believes that it has completely lost contact with the TPS. In order for this code to set the voltage at the TPS, input must have dropped to less than 0.25 volts while the engine was running.

	F	E	D	C	B	A
	Torque Converter Clutch	Serial Data	Check Engine Light	Air Pump Switch	Test	Ground
	Fuel Pump Test	Not Used	Not Used	Not Used	Not Used	Serial Data
	G	H	J	K	L	M

Each terminal of the ALCL connector has a specific purpose. The flange at each end of the ALCL indicates the bottom of the connector, while the flange between terminals C and D indicate the top of the connector. Terminals A and B will always be the two terminals on the upper right.

The TPS receives a 5-volt reference from the ECM along the gray wire CKT 416. If this wire becomes open or grounded, the Code 22 will be set. Additionally, an open or ground in the dark blue wire circuit 417 (the TPS signal wire) can also set this code. A problem in the black ground wire from the TPS could not set a Code 22, however. If the wiring harness, CKT 416, and 417 are not open or grounded, look for a problem with TPS itself.

Code 23

Manifold Air Temperature (MAT) sensor indicating a low temperature, signal voltage high. Code 23 is generated if the MAT detects a temperature less than -31 degrees Fahrenheit while the vehicle is not moving and if the engine has been running more than 4 minutes.

As in the Code 15, this code is generally set when there is an open circuit either in the 5-volt reference wire coming from the ECM, the tan wire CKT 472, or in the black ground wire CKT 452 returning back to the ECM. Should these wires be good (no open circuits), then test the sensor for the proper resistance.

Estimate the temperature of the air in the intake manifold, then use the same chart provided under Code 14. You are looking for an extremely high resistance (in excess of 80,000 ohms) in the sensor for this code to set; therefore your estimate of intake manifold temperature need not be exceptionally accurate unless the ambient temperature where you are working on the car is well below 0 degrees Fahrenheit.

Code 24

Vehicle Speed Sensor (VSS) circuit. The ECM will generate a code relating to the VSS circuit only when the following rather complicated set of conditions are met:
- Vehicle speed indicated is 0 mph
- Engine rpm is between about 1,500 and 3,500
- TPS is less than about 2 percent open
- The transmission is not in park or neutral
- ECM has not previously stored a Code 21, 23, 33, or 34
- All of the above conditions have been met for 4 to 10 seconds

The VSS is generally located on the transmission at the speedometer drive. On applications using a speedo cable it is mounted in the speedometer head.

There are two ways that the ECM receives a signal from the VSS. First as a sine wave, the frequency of which increases as the speed of the vehicle increases. The second way is as a square wave. This is on both transmission-mounted and speedo head-mounted applications that use a buffer circuit. On these, the ECM feeds 12 volts into the buffer. The buffer then grounds and un-grounds this voltage at a frequency proportional to the speed of the vehicle.

The vehicle speed sensor affects the operation of the lock-up converter clutch, idle air control stepper motor and ignition timing (on some applications).

Code 25

Manifold Air Temperature (MAT) sensor indicating a high temperature. The MAT signal wire, CKT 472 (usually tan but sometimes black and pink), is grounded somewhere pulling the signal voltage low.

Code 25 will set if the MAT circuit indicates an air temperature above 258 degrees Fahrenheit for 3 seconds or more. The MAT has a significant effect on only the air-fuel ratio when in open loop; however, either a Code 23 or 25 failure in this circuit can cause some severe driveability symptoms since the ECM will go into the limp-home or back-up mode of operation.

Troubleshooting this circuit is relatively easy since the only causes for the code would be that the CKT 472 (the tan wire) is shorted to ground, or that the MAT sensor itself is shorted. Start the engine and disconnect the MAT connector. Allow the engine to run for about 30 seconds, then shut the engine off and pull trouble codes again. If the Code 23 is now present, the MAT sensor should be replaced. If the ECM does not generate a Code 23 then CKT 472 is shorted to ground.

Code 32

Code 32 refers to a failure in the EGR control system on the 2.8-, 5.0- and 5.7-liter engines. On the 5.0- and 5.7-liter applications, when the ECM commands the EGR valve to open by decreasing the duty cycle from 100 percent, it expects the 12

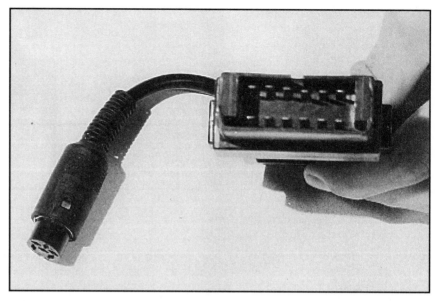

This adapter connects the ALCL to the diagnostic scanner. When you purchase most scanners, you will find connectors compatible with Chrysler, Ford and a 5-pin connector used on the old carbureted cars. Only the 12-pin connector is used on Chevrolet fuel injection.

volts normally on the dark green wire (CKT 999) connected to ECM terminal C15 to drop low. These conditions must first be met, however:

• Coolant temperature is greater than 176 degrees Fahrenheit
• EGR duty cycle command from ECM is greater than 48 percent
• TPS is less than wide-open throttle but greater than idle
• Codes 21, 22, 33, 34 are not present
• All these conditions are met for 4 minutes
• Or the vacuum detecting switch was closed during start-up

When vacuum is applied to the EGR valve, the EGR diagnostic switch detects the additional heat passing through the EGR valve on the Camaro, and through the EGR external tube on the Corvette. It then grounds the 12-volt signal on the dark green (CKT 999) wire connected to terminal C15 of the ECM.

The 2.8-liter engines used in the Camaro are similar, but when the ECM increases the duty cycle to the EGR solenoid allowing vacuum to be applied to the EGR valve, vacuum is also applied to the diagnostic switch. As the switch closes it pulls down the

voltage on the white (CKT 997) wire connected to ECM terminal D8.

The 2.8-liter used in the late-model Celebrity, Beretta and Corsica use an EGR pintle position sensor. This is a potentiometer; its signal wire (pink, CKT 911) is connected to ECM terminal BA3. This sensor not only informs the ECM about whether or not the EGR valve is opening, but also about the exact amount of opening. On these applications Code 32 will be set if the EGR position does not agree with the duty cycle signal being sent to the EGR by the ECM.

TBI applications set a Code 32 when Block Learn cell 10 has a reading greater than 12 different from the closed throttle cell, or when cell 10 exceeds 140. Cell 10 of the 16-cell Block Learn Multiplier is the cell the ECM will be operating when the car is running down the road at 55 mph.

This increase in Block Learn occurs because space in the combustion chamber that would have filled with inert exhaust gases is now filled with air (oxygen). The oxygen sensor detects this extra oxygen and enriches the mixture by increasing the Block Learn. When the Block

CHECK ENGINE

The Check Engine light makes it possible for the driver to know if the ECM has detected fault in one of its circuits. When a fault is detected, the ECM turns on the Check Engine light and the driver knows that there is a code set. When the light goes out, it means that the fault has corrected itself.

Learn exceeds the parameters described in cell 10, Code 32 will be set into the diagnostic memory. For more information about Block Learn, refer to the section on serial data in this chapter.

Code 33

Code 33 is a split code. If the car uses a MAF, then Code 33 indicates an unrealistic flow of air into the intake manifold at a near-closed throttle position. On cars that are not equipped with a MAF, Code 33 relates to an indicated high manifold pressure (low vacuum) from the MAP sensor.

MAF Code 33

Chevrolet uses two different types of MAF sensors. The 5.0- and 5.7-liter Tuned Port Injection applications use the Bosch Hot Wire MAF, while all other applications use the Delco.

On the Bosch MAF, Code 33 will set when the ECM has seen a signal from the MAF indicating an airflow greater than 45 gps (grams per second) (2.2 volts) when the engine is first started, or TPS is less than full throttle and rpm is less than 2,000.

The Delco MAF will generate a Code 33 if the following occurs: In the key on/engine off mode, the indicated airflow exceeds 20 gps; or the engine is running at less than 1,300 rpm, TPS indicates less than 8 percent opening, airflow indicates greater than 20 gps, and all these conditions are met for more than 2 seconds.

This code is often the result of an intermittent condition and will be difficult to troubleshoot. In some cases it may be necessary to take an educated and expensive guess. Check the wiring harness thoroughly. If it checks out okay, look at the routing of the spark plug cables. These cables have a large enough magnetic field around them when the engine is running to induce a pulse into the MAF harness that could be interpreted by the ECM as a high-frequency or high-airflow indication from the MAF. If the routing of the ignition cables is okay and the harness is good, then replace the MAF.

MAP Code 33

Cars equipped with MAP will get a Code 33 when the ECM sees a high manifold pressure (low vacuum) indication from the MAP sensor. Parameters that will typically set a MAP Code 33 are:
• Engine is running
• Manifold pressure indicates greater than 11 psi (about 8 inches of vacuum) with the air conditioner off, and greater than 12.5 psi (about 5 inches of vacuum) with the air conditioner on
• Throttle angle is less than 2 percent
• Above conditions are met for 2 seconds

Whether the car has a MAP or a MAF, Code 33 essentially has the same meaning: the ECM is being told that there is considerably greater air flowing into the intake manifold than either the engine rpm or indicated throttle position would permit. Therefore, as far as the ECM is concerned, this is an illogical reading and a trouble code is set.

Code 34

Code 34 is the opposite of Code 33; however, like Code 33, this code can relate to either the MAF or the MAP. In either case, the code implies that the sensor is reporting to the ECM an impossibly low grams per second of airflow.

Trouble codes, which identify the circuit where a fault has been detected, can be extracted through the Check Engine light. When a jumper wire is placed between terminals A and B of the ALDL connector with the ignition switch on (and engine off), the Check Engine light will flash out trouble codes three times each and repeat them indefinitely. Shown here is a Code 12 (one flash, a brief pause, then two flashes).

MAF Code 34

On the Bosch MAF sensors, Code 34 sets when the airflow is less than 2.5 gps when the engine is first started, or rpm is above 600 and TPS is above 6 percent (this equates to about 2,300 rpm in neutral).

On Delco MAF systems, a Code 34 will set when the engine is running and MAF is disconnected or a signal wire (yellow, CKT 472) is shorted to ground.

In general, a Code 34 means that the ECM believes that there is less grams per second of air entering the engine than there should be, based on the input signals from the other sensors. Items other than a defective MAF that could cause this are false air, vacuum leaks, or coking behind the throttle plates or on the MAF-heated element. An erratic idle can also result in the generation of a Code 34. Inspect all of the tune-up components as well as the IAC valve for wear or damage.

Like the Code 33, this will often be an intermittent problem that may not be occurring as you troubleshoot. Carefully check for the problems listed. Clear the code and if it returns, replace the MAF.

MAP Code 34

On TBI and many PFI applications, Code 34 will refer to a low-voltage signal from the MAP sensor. This means that the ECM is receiving a signal from the MAP indicating an impossibly high manifold vacuum. The code will set when:
- Engine rpm is less than 600
- Manifold pressure reading is less than 13 kPa (kilopascals)
- Conditions are met for 1 second

Or when:
- Engine rpm is greater than 600
- Throttle angle is greater than 20 percent
- Manifold pressure is less than 13 kPa
- Conditions are met for 1 second

If the car you are working on has both a MAF and a MAP, Codes 33 and 34 will refer to the MAF, while 63 and 64 will refer to the MAP sensor.

Code 35

Code 35 refers to Idle Speed Error. This code will be generated whenever the actual idle speed is 300 rpm above or below the desired idle speed for 50 seconds.

Although this code can be caused by a faulty IAC motor, it is far more likely to be set as a result of one of the following conditions:
- The engine is running too rich or too lean (refer to codes 44 and 45)
- There is evidence that the control valve pintle is dragging the bore of the IAC passage
- Improper air conditioner compressor clutch operation
- Hunting idle

If none of these conditions is present, then with the engine idling, connect a test light to ground, disconnect the IAC electrical connector and probe all four of the IAC terminals. The test light should blink on and off. If it does, replace the IAC motor. If a circuit does not blink, check for an open, short or short-to-ground in the circuit and repair as necessary. If the wire that is not blinking does not have a short, ground or open, check the connector at the

ECM for tightness and corrosion. If this checks out, replace the ECM.

Note: Because damage to the ECM can be caused by an internal short in the IAC coils, check the resistance in these coils before installing a new ECM or reinstalling the old ECM. The resistance in the IAC coils should be greater than 20 ohms.

Code 36

Code 36 relates only to MAF-equipped TPI applications. The code indicates that the ECM has failed to detect 12 volts available on the black wire connected to terminal D12 (CKT 900) when burn-off relay is not energized. Since the relay is energized by being grounded through terminal D12, the ECM expects to see 12 volts when the MAF is not in the burn-off mode. Code 36 is most often set as a result of an open circuit in the black wire connected to terminal D12 of the ECM, or in the orange wire running from the burn-off relay through a fusible link to the positive terminal of the battery.

Code 41

In an effort to reduce the total number of ECMs that Chevrolet uses in its fleet, one basic programming may be used with several different engine applications. The ECM is made aware of which engine it has been teamed up with by means of the PROM or CALPAK. If the ECM memory contains a Code 41, it will be necessary to replace the CALPAK. Occasionally there is nothing wrong with the CALPAK and installation of the new one will not do away with the Code 41. If the code recurs, then check the CALPAK connections in the ECM. If they are okay, replace the ECM.

Code 42

Code 42 is set when the engine is running at greater than 600 rpm and there has been no pulse on the white wire from the ECM to the ignition module (CKT 423) for more than 200 milliseconds (ms). (At

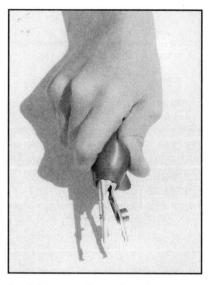

Special alligator clips such as this are used to penetrate wires in order to sample voltages without damaging the insulation.

2,000 rpm, 200 ms represents about 27 spark plug firings.) Code 42 can also be set if the voltage on the tan and black wire (CKT 424) remains low (near 0) or goes low while the car is being driven. The ignition module is designed to ground out the EST (Electronic Spark Timing) signal that travels down the white (CKT 423) wire unless there is 5 volts on the tan and black wire. This locks the timing at a fixed position while the engine is being started.

Common causes of a Code 42 include bad connections, frayed or broken wires, and rubbed-through insulation on either the tan and black, white, or purple and white (CKT 430) wires running between the ignition module and the ECM.

For more on troubleshooting this problem, refer to chapter six on tune-ups.

Code 43

Code 43 relates to the detonation sensor system. Chevrolet uses two types of detonation (knock) sensors.

The first type dates back to the 1970s and uses an interface module between the sensor and the ECM. As of this writing this is the most common style, being used on TPI applications, as well as most of the PFI and

CHECK ENGINE

Code 13 relates to the oxygen sensor. Basically, what causes a Code 13 to be set is a perceived open circuit in the purple lead to the oxygen sensor. The code might also occur if the oxygen sensor is contaminated or sooted-up.

TBI applications. The interface module, known as an Electronic Spark Control (ESC) module, has five pins but only four of them are used.

• Pin A: Unused
• Pin B: CKT 439, a pink and black wire carrying ignition power to both the ESC module and the ECM
• Pin C: CKT 485, a black wire that carries the ESC signal to ECM
• Pin D: CKT 486, a brown wire that is grounded to the engine to provide a power ground for the ESC module
• Pin E: CKT 496, the white wire that carries the knock sensor's signal to the ESC

On the cars that use the ESC module, Code 43 will set when the ECM sees the voltage on terminal B7—the black ESC signal wire—drop low for more than 5 seconds with the engine running. A Code 43 might also be set if the ESC system fails the functional test. The functional test is performed by the ECM once per start-up. When the coolant temperature reaches about 194 degrees Fahrenheit and the throttle is opened to nearly wide open the ECM will begin to advance the timing, deliberately allowing the engine to ping in order to test whether or not the ESC will detect the detonation. If no detonation is detected then a Code 43 will be set into the ECM's memory.

The second type of detonation sensor system is used on the Generation II 2.8-liter engine and others. This system does not use an interface module; the knock sensor feeds directly into the ECM. Code 43 will be set if the voltage on the dark blue and white wire connected to ECM terminal YFR9 drops below 1.5 volts or rises above 3.5 volts for more than half a second. When troubleshooting this circuit, keep in mind that the ECM maintains a 2.5 volt signal on the wire whenever the engine is not knocking.

Code 44

Code 44 is set by a perceived lean exhaust condition. This lean running condition might be a result of air-fuel ratio errors, or it might be a result of:
• A grounded oxygen sensor wire
• Restricted injector(s)
• Contaminated fuel
• Low fuel pressure
• Exhaust leaks
• Leaking air switching valve in the air pump system

Code 44 sets when the voltage output of the oxygen sensor drops below 0.2 volts and remains there for more than 60 seconds while the system is operating in closed loop.

Code 45

Code 45 occurs when the oxygen sensor indicates a rich exhaust condition. Code 45 will set when the oxygen sensor output remains over 0.7 volts for more than 30 seconds, providing that the throttle angle is between 3 percent and 45 percent and the engine has been running for at least 1 minute.

This code can be set by a short to voltage in the oxygen sensor circuit, a lead-contaminated oxygen sensor or any of the following:
• High fuel pressure
• An injector that is either seating slowly or leaking
• Induced voltage (current) from the secondary ignition system
• The evaporative control system is either saturated or the purge valve is defective
• An erratic TPS signal
• The EGR valve is not seating properly

Code 46

Code 46 is unique to the Corvette. It refers to the Vehicle Anti-Theft System (VATS). Code 46 problems should be referred to the dealer and are not dealt with in this book.

Code 51

What follows are the 50-series codes. These codes usually indicate a detected fault that plays a major role in ECM function. Therefore, if any 50s codes are obtained when the codes are pulled, they should be diagnosed first.

Code 51 will be set only if there is a defective MEMCAL, defective connections to the MEMCAL or a defective ECM. The MEMCAL provides the ECM with data specific to the application in which the ECM is installed, and provides limp-home information to replace the signals from a defective sensor.

Code 52

CALPAK error. Code 52 is set whenever the ECM does not recognize the right information coming from the application calibration package known as the CALPAK. The causes of this code include a defective MEMCAL unit, bad connections of the MEMCAL or a defective ECM. The CALPAK is the portion of the MEMCAL that provides the limp-home mode information.

Code 53

System over voltage. When Code 53 is found in the diagnostic memory, it means that the ECM has seen more than 17.1 volts at the ECM ignition input pin for more than 2 seconds. Diagnose and repair the charging system as necessary.

Code 54

CKT 120 to the fuel pump is monitored by the ECM. Code 54 will set when the ECM detects less than 2 volts going to the fuel pump within 1.5 seconds of the last reference pulse.

Code 14 is usually caused by either a short in the coolant temperature sensor or a grounded yellow wire to the CTS. A Code 15 is usually caused by an open in the yellow wire, the black wire, or the CTS.

Code 55

Replace ECM. No ifs, ands, or buts—replace it!

Code 61

Degraded oxygen sensor. This code results when the response time of the oxygen sensor drops dramatically. This condition is usually the result of an oxygen sensor that has been contaminated with RTV silicone or leaded gas.

Code 63

Refers to the MAP sensor section of Code 33. Code 63 is used to identify a low-voltage signal from a MAP sensor on cars that use both a MAP and a MAF.

Code 64

Refers to the MAP sensor section of Code 34. Code 64 is used to identify a high-voltage signal from a MAP sensor on cars that use both a MAP and a MAF.

Serial Data

One of the nicest things General Motors and Chevrolet have done for the professional technician has been to provide serial data information from the ECM. Serial data tells the technician what the computer's innermost thoughts are: it provides information from the input sensors as well as what decisions the computer is making based on its sensor reading.

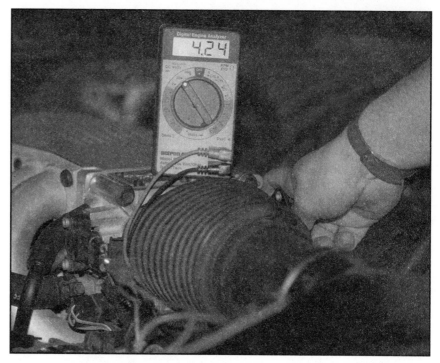

Code 21 is set when the ECM sees a voltage from the TPS that is much higher than it could logically be with the other current engine running conditions. For example, the ECM has been programmed to know that the TPS voltage cannot be 4.24 volts for more than 1 to 2 seconds if the engine rpm is only 800.

Scanners

The unfortunate thing about serial data is that in order to obtain the information on the Chevrolet applications, it is necessary to use a diagnostic scanner. These scanners can run well over a thousand dollars, but for the enthusiast or do-it-yourselfer who owns a personal computer, there is a relatively inexpensive alternative. Several companies, including Rinda Technologies of Chicago, have systems or programs available that use a communications link from the car's ALCL connector to a personal computer, allowing the PC to interpret and display ECM data. Where a good scanner may cost $1,200 or more, these programs are available for as little as $295.

To use the scanner, connect it to the ALCL connector, which is located above the driver's right knee on most applications. What follows is a brief explanation of each piece of data displayed, the normal range of readings for each, and some things that might cause abnormal readings.

The ranges given as normal are representative of normal readings, and although they will be reasonably accurate regardless of the application, they may not be precise for all Chevrolet systems. However, used in conjunction with diagnostic techniques such as symptom detection and trouble codes, serial data can be a powerful tool in troubleshooting fuel injection problems.

All methods have their limitations, though. The limitation of serial data troubleshooting is the fact that the information displayed on the scanner represents only samples of the sensor and actuator readings occurring 1.25 seconds apart. What happens in the way of highly intermittent defects occurring between scan updates is lost.

When setting up the scanner to troubleshoot a car, it may need to know the GM division (Chevrolet), which on 1981 and later models is the third digit of the Vehicle Identification Number (VIN), the car line, and the engine type, which is the

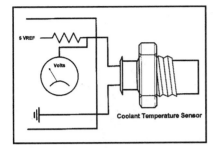

A defect in the coolant temperature sensor will cause either Code 14 or Code 15. If a Code 14 is set, the ECM's internal voltmeter is measuring close to 0 volts on the yellow wire to the coolant temperature sensor. A Code 15 means close to 5 volts on the wire. Codes are set due to an inappropriate voltage.

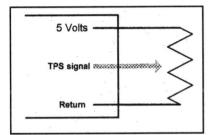

The throttle position sensor can set a Code 21 or 22. When a Code 21 is present, it means that the ECM is receiving a voltage higher than should ever be present on the TPS signal wire. Code 22 indicates that the TPS signal wire is sending 0 volts to the ECM.

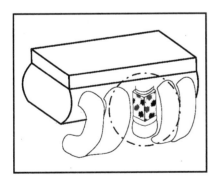

Codes 23 and 25 relate to the Manifold Air Temperature sensor, or MAT. On TPI engines, the MAT is located under the intake manifold between the runners. A Code 23 indicates an open in the MAT circuit, while a 25 indicates that the tan wire to the MAT is grounded or the MAT itself has an internal short.

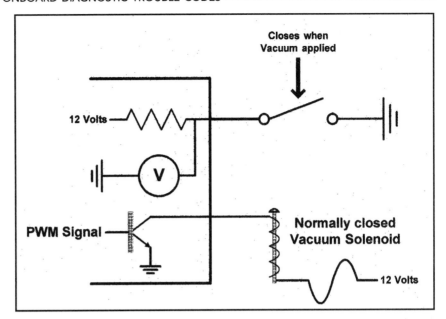

The EGR valve on the 2.8-, 5.0- and 5.7-liter engines detects when vacuum is being applied to the EGR valve. The detecting switch on the 2.8-liter is controlled by vacuum as shown, while the one on the 5.0- and 5.7-liter is controlled by heat through the EGR valve. If the ECM does not detect a drop in voltage on the EGR diagnostic switch at the appropriate time, then a diagnostic Code 32 will be set.

eighth digit. Be ready with this information in case you need it.

Engine Speed

The engine speed data field indicates what the current engine speed is in rpm. The report changes in increments of 25 and should always be within 50 of the desired idle speed on cars that display desired idle. On cars where desired idle speed is not displayed, the engine speed may be as high as 1,500 rpm on a cold engine down to as little as 650 after the engine is warmed up. As a rule of thumb a V-8 will idle about 650–750 rpm warm, a V-6 750–850 rpm and a four-cylinder between 750 and 1,000 rpm.

Inappropriate idle speed can be caused by:
- Sticking or binding throttle linkage
- Incorrect Minimum Air adjustment
- Vacuum leak
- Defective Idle Air Control valve or circuitry
- Damaged or defective motor mounts
- ECM targeting incorrect idle speed

Desired Idle Speed

Desired idle speed is the engine speed that the ECM has targeted as appropriate for current engine operating conditions. The ECM takes into account coolant temperature and engine load factors such as the gear the transmission is in, the pressure in the power-steering system, and the status of the air conditioner compressor clutch to calculate a desired target idle speed.

Not all applications will transmit this data to the scanner. If the target rpm seems inappropriate for the temperature of the car being tested, check the related sensory inputs already listed.

Vehicle Speed

Vehicle speed is the result of the input from the Vehicle Speed Sensor (VSS). Since most of the time serial data is read with the car stationary, the most often read mph reading will be 0. As the car is driven, the MPH reading on the speedometer and the scanner should agree within just a few units.

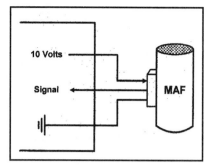

Codes 33 and 34 relate to the sensors that measure the flow of air into the intake manifold. On the TBI and later 2.8- and 3.1-liter PFI applications, this job is done by a MAP sensor. Other PFI and TPI applications use the MAF sensor, as shown above. Code 33 means that an excessive airflow is being reported by the MAP or MAF. Code 34 indicates that there is an impossibly low airflow being indicated.

The EGR control solenoid is used on the 2.8-liter and Tuned Port Injection applications. A Code 32 indicates that the EGR valve is not being controlled properly. The problem could be as simple as a dirty vent filter (shown), or it could be electrical.

Should the vehicle speed read something seriously incorrect, and yet no trouble code has been set, likely causes would be a dirty optical sensor in the speedo head, a defective VSS, or the incorrect GM division punched into the scanner. See Code 24.

Coolant Temperature

The coolant temperature reading comes from the coolant temperature sensor and reads between -40 and 302 degrees Fahrenheit. Practical knowledge of engine operating temperature would be helpful in the proper interpretation of this information. The first start-up of the day should yield a displayed temperature very close to the ambient air temperature. As the engine runs, the temperature should steadily increase to between 195 and 220 degrees Fahrenheit.

Abnormal temperature readings might be the result of a faulty thermostat, a restricted radiator, restricted airflow through the radiator, a defect in the air conditioning condenser or a blown head gasket or other engine defects. Or there could be a problem with the Coolant Temperature Sensor circuit (see Code 14 or 15).

Start-Up Coolant Temperature

The start-up coolant temperature reflects the first coolant temperature detected by the ECM as the ignition switch is turned on during a start-up. If it is the first time the car is started during the day, then the temperature indicated should be close to the air temperature.

Manifold Air Temperature

Manifold air temperature, reported by the MAT sensor, will vary depending on the ambient temperature, the rate of airflow into the engine, the underhood temperature, the engine temperature, and the location of the MAT. If the MAT is in the intake manifold, the indicated temperature will be higher when the engine is at operating temperature than if it is located in the air cleaner. The indicated temperature from a MAT sensor in the intake manifold should drop noticeably when the engine is revved and the in-flow of the air increases.

A typical temperature for this field on a warm engine would be about 95 to 175 degrees Fahrenheit on applications where the MAT is

On engines equipped with a MAP sensor, code 33 indicates that the MAP signal voltage is well above the expected normal range. This could be caused by a broken vacuum line, a defective MAP sensor or the signal being shorted to a high voltage. Code 34 will set as a result of a defective MAP, a broken MAP signal wire, a grounded MAP signal wire or an open in the 5-volt reference.

located in the intake manifold. Incorrect temperature could be the result of an overheating engine, an extremely high ambient temperature, or a faulty MAT sensor or circuit (see Code 23 or 25).

Throttle Position Voltage

The throttle position voltage field shows the voltage output of the Throttle Position Sensor. The volt-

At an idle the MAP sensor should produce a signal of about 1.5 volts. This will vary a little with altitude, weather and engine condition. The signal of 1.94 volts shown above either indicates that the altitude is rather high, that the barometric pressure is extremely low, or that the engine is worn. This voltage, even though it is a little out of the expected range, could never set a trouble code.

The white EST wire is monitored by the ECM and will set a Code 42 if the EST signal is not present. Causes of the code being set are an open in this tan wire with a black stripe, a defective ignition module, or an open, short or ground in either the tan and black by-pass or the white EST wire.

age output of the TPS at closed throttle should be about 0.5 volts (refer to the Minimum Air and TPS Specifications Chart in the appendices) and gradually increase in a linear fashion (within the limitations of scan updating) until wide-open throttle indicates a voltage greater than 4 volts.

An incorrect closed-throttle TPS reading can result from improperly adjusted Minimum Air, improperly adjusted TPS, or a defective TPS or wiring from the TPS to the ECM (see codes 21 and 22).

Filtered Load Variable

Of all of the serial data fields, Filtered Load Variable is probably the most difficult to fully understand, but fortunately is not critical for most diagnoses. The Filtered Load Variable is displayed in grams per cylinder and is used by the ECM to indicate engine load on applications not equipped with a MAP sensor. At an idle, a typical display in this category would be about 50 to 80 grams per cylinder; as the load on the engine increases, the load variable will also increase. Switch on the air conditioner or load the power steering and the number should increase a little.

Oxygen Sensor Voltage

The oxygen sensor voltage category shows what the voltage output of the oxygen sensor was during the last scan. This voltage should be constantly changing and switching back and forth across the crossover point of 450 millivolts (0.450 volts). A voltage higher than 450 millivolts indicates that the oxygen sensor is detecting low oxygen content in the exhaust or a rich exhaust condition. An oxygen sensor voltage less than 450 millivolts indicates that the exhaust has a high oxygen content or lean condition.

An oxygen sensor voltage that does not change or is sluggish indicates that it has been contaminated. If the voltage is pegged on the low end of the range, look for the oxygen signal wire being shorted to ground. If the voltage is pegged on the high end of the scale, inspect the oxygen sensor wire for a short to voltage. If the voltage remains at or very near 450 millivolts, look for an open in the oxygen sensor wire. Refer to Codes 13, 44 and 45 for more information.

This is the ESC module (shown with one mounting wing broken off). The ESC module monitors the detonation sensor and reports detonations to the ECM so that the ECM can retard the timing. A defect in this system will set a Code 43.

Idle Air Position

The range of the idle air position field is 0 to 255. The number represents the position to which the ECM believes it has moved the Idle Air Control valve. The normal position for this stepper motor should be between 5 and 45 at an idle. The higher the number is, the more air the ECM is trying to let into the intake manifold to increase the idle speed. The lower the number, the more the ECM is trying to lower the idle speed.

An IAC number lower than 5 indicates that there may be air entering the

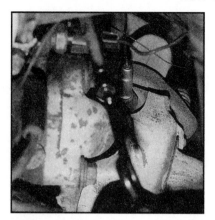

A Code 44 will set when the oxygen sensor voltage is extremely low for an extended period. This could result from a defective sensor, a grounded purple wire between the sensor and the ECM, or excessive amounts of oxygen in the exhaust system. The Code 45 indicates that the engine is running rich, a very low oxygen content in the exhaust. This code would also be set if the oxygen sensor signal wire were shorted to voltage.

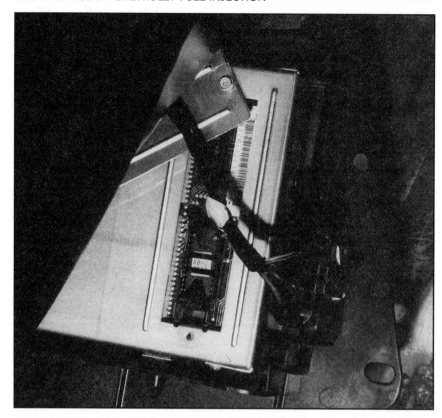

Codes 51 and 52 relate to the CALPAK or PROM located below the access cover in the ECM. The CALPAK is the only ECM microprocessor that is serviced separately from the entire ECM.

Code 55 means replace the ECM. Be sure and test the actuator circuits suggested in the instructions of the new ECM. The reason for the ECM failure could be a shorted actuator circuit.

engine through a vacuum leak or that the Minimum Air adjustment is incorrect. In some cases, a defective (jammed) IAC valve could cause this reading.

An excessively high reading indicates that the engine is under an abnormal load, such as torque converter problems. It may also mean that the ECM is trying to increase the idling speed of the engine as a response to a perceived need for more output from the alternator.

With key on and engine off, the IAC position will be between 100 and 150. This will permit enough airflow into the engine to allow easy starting with the driver's foot off of the gas pedal. Also, as the car is driven, the IAC will be placed in this position to act as a dashpot during deceleration.

Airflow Rate

The Airflow Rate reading is in grams per second of airflow. For most idling engines that use this sensor, the display on the scanner should read about 5 to 10 gps. The reading should increase as the speed of the engine increases.

A low gps reading at an idle (less than 5) indicates that the engine is getting air from a source other than through the MAF. This false air could be in the form of a vacuum leak or a poorly fitted connecting tube between the MAF and the throttle assembly.

A gps reading that is too high indicates a malfunction in the MAF itself. Refer to Code 33 for diagnosis.

With the ignition switch on but the engine not running, the Delco MAF sensor will read about 2 or 3 gps. The reason airflow is indicated even when there is no air entering the engine is that the heated element will give up some of its heat to the air that is laying static around it. This loss of heat causes the MAF to indicate that there is air flowing into the engine. Under these conditions the Bosch MAF sensor readings are erratic. Malfunctions in the MAF are monitored by Codes 33 and 34.

MAP Sensor Volts

Perhaps more than any other serial data field, the logic of the information provided in this field is easiest to see in MAP volts. The MAP indicates the amount of pressure in the

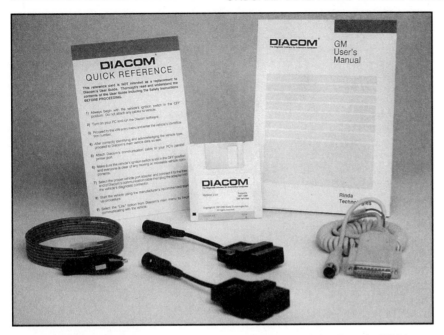

Owners of personal computers can buy a program to extract serial data from the ECM for only a few hundred dollars.

intake manifold. This might be a little confusing since we are normally accustomed to thinking in terms of a vacuum in the intake manifold.

When the engine is not running, the pistons are not moving up and down in the cylinders to create a vacuum, thus the pressure in the manifold is the same as atmospheric pressure. At sea level, normal atmospheric pressure is considered 29.92 inches of mercury (in. Hg.). At this point the MAP sensor field will display something in the area of 4.5 volts. This reading may vary depending on altitude, weather and from one vehicle to another.

At an idle, the manifold vacuum reading will be between 18 and 20 inches of vacuum. If you subtract the idle vacuum reading from the atmospheric pressure reading you will end up with about 10 inches of mercury pressure in the manifold at an idle. The 10 inches of pressure is about one-third of the atmospheric pressure. The MAP voltage at an idle is about one-third of the MAP voltage with the key on and engine off, or 1.5 volts.

Some scanners or applications will display this field in kilopascals (kPa). For those applications, the key-on reading should be around 100 kPa and about 35 kPa at an idle.

Abnormal readings could be caused by vacuum leaks, a worn engine, an engine running too rich or too lean, or an ignition that misfires.

On applications that do not use a MAF, the MAP is monitored by Code 33 for an excessively high pressure (low vacuum) signal, and Code 34 for a low-pressure signal. On cars that are also equipped with a MAF, such as many late-model 2.8-liter engines, the codes for the MAP are 63 and 64.

Injector Pulse Width

Injector pulse width is a measurement of the length of time that the injectors are open measured in milliseconds (ms). For most Chevys this should read about 0.7 to 1.5 ms at an idle. If the engine speed is slowly increased without moving the car, the pulse width should not increase significantly. This is because the engine's need for fuel per firing of each cylinder does not increase dramatically until the load on the engine is increased. Accelerating the engine rapidly, even with no load on it, should cause a marked increase in pulse width.

As a diagnostic aid, the most helpful thing about this field is that as the injectors become restricted, the pulse width will increase. Injector pulse width alone should never be used to isolate the cause of a driveability problem. Other important factors to consider would be the Integrator, the Block Learn Multiplier, fuel pressure and oxygen sensor readings.

If the injector pulse width is not displayed in serial data, use the formulas described in the tool chapter to calculate frequency and duty cycle of the injector pulse. Use the following formula to calculate injector pulse width:

$$\frac{1}{Frequency} \times \frac{Duty\ cycle}{100}$$

Spark Advance

Spark advance is an indication of the number of degrees of advance the EST section of the ECM has added to initial timing. It is not unusual to see this field indicating 25 to 35 degrees of advance at an idle.

Knock Retard

When the ESC module receives a signal from the knock sensor indicating that a detonation has been sensed, it sends a signal to the ECM requesting that the timing be retarded. This knock retard field indicates how many degrees the timing has been retarded from its spark advance setting to control detonation. Because a slight amount of pinging is to be considered normal on applications equipped with the ECS system, a reading of four or so should not be interpreted as a problem.

Spark Control Counts

The range for the spark control counts data field is between 0 and 255. It is a count of how many times the ECM has responded to a knock detected by the ESC system. The number displayed is irrelevant in troubleshooting, however. The speed that the number increases might be indicative of excessive pinging caused by vacuum leaks, lean mixture, a defective EGR system, or over-advanced initial timing.

Integrator

When the engine is running in the closed-loop mode, the Integrator is responding to signals from the oxygen sensor to control fuel delivery through the injectors. The total range of this field is from 0 to 255. A reading of 128± (some recent literature uses a tolerance of 10) should be considered within the normal range. The Integrator responds quickly to oxygen sensor readings.

If the reading is higher than the normal range, it indicates that the ECM is detecting a lean reading from the oxygen sensor and is increasing the injector pulse width to compensate. The number displayed is not a measurement of time, volume, or anything concrete; rather, it is a numbering system designed to represent trends in the variation from normal of the air-fuel ratio. A high Integrator reading might be caused by a vacuum leak, low fuel pressure, restricted injectors, an exhaust leak, an ignition misfire, low compression, or a defective oxygen sensor. See Code 44 for more details.

If the reading is lower than the normal range, it indicates that the ECM is trying to lean out the air-fuel ratio by decreasing the injector pulse width. Low Integrator readings can be caused by the following: A defect in the evaporative canister system; high fuel pressure; a leaking injector or a leaking cold-start injector (5.0-, 5.7-, and some 2.8-liter PFI applications only). See Code 45 for more details.

Anytime the ECM is operating in open loop, the Integrator will be steady at 128.

Block Learn Multiplier

Unlike the Integrator, the Block Learn Multiplier is stored in memory and can adjust to compensate for air-fuel ratio problems even when the engine is operating in open loop. Imagine the Block Learn attached to the Integrator by an elastic cord. When the Integrator makes a sudden movement, it will stretch the elastic; if the Integrator remains where it has moved to, the Block Learn will soon

Voltmeter reading taken directly at the sensor wiring harness should be the same as readings taken with a diagnostic scanner. Variations between the two indicate wiring problems between the ECM and the sensor.

A low grams per second airflow rate at idle could be an indication of false air (a vacuum leak). Inspect the rubber connecting boot between the MAF and the throttle assembly.

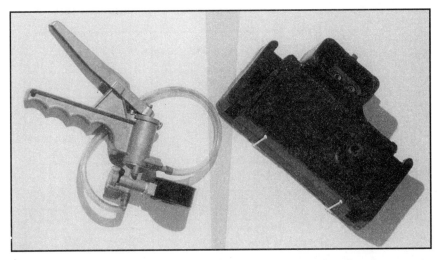

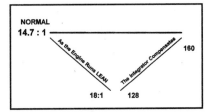

Connect a hand-held vacuum pump to the MAP sensor and monitor the voltage as you apply and release the vacuum. The voltage should drop smoothly as the vacuum is being applied, and rise smoothly as the vacuum is being released.

One of the most important data fields is the Integrator, which indicates how the ECM is correcting for errors in the air-fuel ratio. The number displayed in the Integrator data field increases as the engine leans out. A vacuum leak, restricted injectors, or low fuel pressure are common causes of an elevated Integrator. If the fuel pressure is too high or an injector is leaking, the Integrator will drop.

be snatched toward it. It could be accurately said that the Integrator responds to the oxygen sensor and the Block Learn responds to the Integrator.

At each start-up, the Block Learn will be programmed with the readings it had the last time the engine ran. As a result, a repair made while the engine is shut off may cause a driveability problem until the Block Learn reprograms to the repaired engine.

High and low BLM readings are caused by the same conditions that cause high and low readings in the Integrator.

Block Learn Cell Number

Because vacuum leaks, leaking injectors and such will affect the air-fuel ratio to a different degree at different engine speeds and loads, the Block Learn is divided into 16 cells numbered 0 to 15. Each cell represents a different rpm range. Refer to the diagrams to see how Block Learn progresses from one cell to another.

Miscellaneous: Battery Voltage

The battery voltage field indicates the perceived voltage from the battery to the ECM in increments of 0.1 volt.

Fuel Pump Relay Voltage

Fuel pump relay voltage is the voltage available to the fuel pump. See Code 54 for more details.

Engine Run Time

Engine run time is the length of time the engine has been running, displayed in minutes and seconds.

Cooling Fan Duty Cycle

The electric radiator cooling fan is controlled by a variable duty cycle on some applications. This field indicates the fan speed that the ECM is requesting. Zero percent is fan off, while 100 percent is fan at full speed.

Canister Purge Solenoid

Canister purge solenoid field indicates whether the ECM is purging the evaporative canister. For most applications this is a simple on-off signal. For some applications, such as late-model TPI Corvettes, this field is called Canister Purge Duty Cycle and is a percentage reading from 0 percent with the canister purge valve closed, to 100 percent with the purge valve fully opened.

EGR Duty Cycle

Some applications, such as the 2.8-, 5.0- and 5.7-liter engines, control the EGR valve by controlling a pulse to the EGR vacuum control solenoid. On the 2.8-liter, a reading of 0 percent indicates that the EGR valve is being requested to remain closed. On the 5.0- and 5.7-liter, the closed signal is 100 percent.

EGR Diagnostic Switch

When vacuum is applied to the EGR valve, the pintle will be lifted. At the same time either a temperature switch or a vacuum switch will detect that the EGR has had vacuum applied, and this field will toggle from off to on.

If the ECM's attempt to apply vacuum fails to open the EGR valve either because of a vacuum leak or other defect, then this field will fail to toggle. See Code 32 for more details.

Loop Status

The indication in the loop status is either open or closed. It simply reports whether the ECM is operating in the open- or closed-loop mode.

Oxygen Status

The oxygen status field indicates the general trend of the oxygen sensor reading. If the exhaust gases are mostly indicating a rich condition, that is what the field will display.

12	13	14	15
08	09	10	11
04	05	06	07
00	01	02	03

The Block Learn Multiplier (BLM) is similar to the Integrator, in that it corrects for fuel-injection errors. Unlike the Integrator, Block Learn corrections are stored in a memory. Since fuel-correction requirements might be different under different engine load and rpm conditions, the Block Learn is changed in 16 separately programmable memory cells relating to engine load and rpm. Only the cell corresponding to the current engine conditions will be displayed on the scanner.

LOAD ↑			
12 128	13 128	14 128	15 128
08 128	09 128	10 128	11 128
04 128	05 128	06 128	07 128
00 128	01 128	02 128	03 128

RPM →

When the fuel injection system is putting the correct air-fuel ratio into the engine, all BLM cells will be programmed at 128.

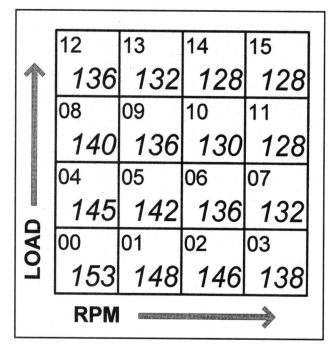

LOAD ↑			
12 136	13 132	14 128	15 128
08 140	09 136	10 130	11 128
04 145	05 142	06 136	07 132
00 153	01 148	02 146	03 138

RPM →

When the engine is running lean, the Block Learn numbers will increase. These numbers reflect how the BLM might reprogram in response to a medium-size vacuum leak. Note that the need for additional fuel decreases as the engine load and rpm increase.

LOAD ↑			
12 105	13 115	14 121	15 128
08 98	09 109	10 118	11 122
04 94	05 98	06 110	07 115
00 90	01 92	02 100	03 109

RPM →

When the engine is running rich, the Block Learn numbers will decrease. These numbers reflect how the BLM might reprogram in response to a leaking injector. Note that here again the requirement for correction decreases as the engine load and rpm increase.

Park-Neutral Switch

The park-neutral switch display should read "Off" whenever the transmission is in park or neutral and "On" in any other gear.

Power-Steering Switch

When the pressure in the power-steering system increases as the steering approaches full lock, the field will toggle from "Off" to "On," telling the ECM that the engine is under an extra load from the power steering.

This information is used by the ECM as a need to increase the opening of the IAC to maintain a steady idle speed in spite of the power-steering load.

Gear Switches

The gear switches fields represent the status of various gears as the car is being driven. This display may read either "On" or "Off" when not in the indicated gear. What is important is that the display toggles as the car goes in and out of the gear.

High-Battery Indicator

High-battery indicator will display "Off" unless the charging system voltage approaches 16.9 volts at which point it will toggle to "On."

Cooling Fan Request

"On" indicates that the radiator cooling fan has been requested by the ECM.

Skip Shift Enabled

Skip shift enabled applies only to the Corvettes equipped with the six-speed manual transmission. Under casual driving conditions the field will toggle to "Yes" and require the driver to alter the normal shifting pattern by skipping second and third gears.

Skip Shift Active

"Yes" will be displayed on cars equipped with the six-speed transmission when second and third gears have been locked out.

Learn Control

There are a few applications that present the learn control field. An "On" display indicates that the ECM is in closed loop and that the Block Learn is ready to function.

Vehicle Anti-Theft System (VATS)

The display in the Vehicle Anti-Theft System data field will either indicate "Pass" or "Fail." If fail is indicated, it means that the VATS has failed to deliver permission to the ECM for the injectors to work, and the car will not start. Typically a fail signal results from the wrong key being inserted into the ignition.

Air Divert Solenoid

"On" indicates that the air from the air pump is being routed to the air switching valve. "Off" indicates that the air is being routed to the air cleaner.

Air Switch Solenoid

"On" indicates that air pump air is being directed to the exhaust manifold. "Off" indicates that the air is being pumped to the catalytic converter.

Shift Light

The shift light field should read "Off" unless the ECM feels that the manual transmission should be shifted to the next highest gear to maintain good fuel economy.

Air Conditioner Clutch

The air conditioner clutch display toggles to "On" when the ECM has directed the air conditioner compressor clutch to engage.

Air Conditioner Request

When the display toggles to "On," it indicates that the automatic climate control system has requested the ECM to turn on the air conditioner compressor clutch.

Summary

This covers the bulk of the serial data fields delivered to a scanner by the ECM. It should be noted that not all applications will transmit all of this data, and that there may be data fields delivered on a given application that is not discussed here.

Using serial data along with symptoms and trouble codes makes for a well-rounded diagnostic approach that will solve the bulk of driveability problems.

CONTENTS

VISUAL
INSPECTION   108

VACUUM HOSES   109

WIRING HARNESS
AND ROUTING   109

FUEL LEAKS   109

EXHAUST LEAKS 109

SCOPE ANALYSIS 109

CODE PULLING   111

FUEL PRESSURE
CHECK   111

FLOW CHART
USAGE   111

SENSOR
CIRCUITS   111

SENSOR CIRCUITS
REFERENCE   112

ACTUATORS   115

YOU CAN DO IT! 115

CHAPTER NINE

HOW TO USE FAULT CODES

Today's cars have turned yesterday's art of troubleshooting rough idling, hesitations and stalling into a precise science. Over the years since the introduction of electronic fuel injection, the auto repair industry has gained ever-increasing knowledge of how EFI works and how to repair it. To a large extent, these systems have caused the do-it-yourselfer—the car owner with skills capable of doing a tune-up or repairing a four-barrel Holley—to find himself or herself completely lost and without the basic skills necessary to solve relatively simple driveability problems. This section begins with a look at a logical diagnostic sequence, then provides the detail to put this sequence to use.

Any logical troubleshooting sequence must begin by looking at the simplest and least expensive solution, and then proceed to the most complicated and most expensive. Realistically there are several different ways a logical progression might be organized. It is even possible that from one car to another, one diagnostic situation to another, the sequence might be altered.

We will approach this troubleshooting sequence by addressing the following issues: Visual Inspection; Scope Analysis; Code Pulling; Fuel Pressure Check; Flow Chart Usage; Driveability; and Serial Data.

Visual Inspection

A thorough visual inspection is probably the single most important diagnostic procedure that a technician can perform in finding a driveability problem. As a former trainer with a major scope manufacturer, many is the time that I have seen the human eyeball outperform $25,000 engine analyzers.

During a visual inspection, the following items should be carefully examined.

Coolant Level and Mix

Coolant used to be called antifreeze. In the old days you could get by without it in climates where oranges grow. Today, anti-freeze is an assumed component of the computerized engine control system. Engines run at temperatures that can exceed 230 degrees Fahrenheit. At these temperatures water will boil, even in a pressurized system. Thus, anti-freeze is necessary to prevent boil-over. Anti-freeze also has the ability to help spread the engine's retained heat more evenly throughout the block and head. Pure water will allow localized hot spots to build up, which can contribute to detonation.

As bad as pure water is for the operation of the engine, pure anti-freeze can be worse. The principal element of anti-freeze is ethylene glycol. The freezing point of water is 32 degrees Fahrenheit. The freezing point of ethylene glycol is only slightly below that. It is only when the water and ethylene glycol are mixed that the freezing temperature is significantly lowered. A mix of about 40-percent water and 60-percent ethylene glycol will lower the freezing temperature of the mixture to -65 degrees Fahrenheit.

Today's engine analyzers follow computerized diagnostic routine to quickly diagnose basic engine, ignition, and fuel problems. A good place to begin a diagnosis would be to take the car to a shop that has one of these machines.

Besides the obvious potential of overheating the engine, a car that is low on coolant may run rich, get poor fuel economy and have detonation problems.

Radiator Hoses, Connections, and Clamps

Although defective radiator hoses, clamps, or connections will not cause a driveability problem, they can cause a loss of coolant, which, in turn, can cause any of the problems already described.

Battery Cables

Incorrect voltage to the ECM or the fuel pump can cause improper operation of those components. Good connections are essential for delivering the proper amount of voltage to all of the vehicle components. This is especially true as the load on the electrical system increases when lights, wipers, and rear-window defoggers are turned on.

Inspect battery cables for signs of bubbling or swelling under the insulation near the cable end. This is a tell-tale sign of hidden corrosion.

Important: Do not remove the battery cables at this point! Doing so will erase trouble codes that may be important to the proper diagnosis of the driveability problem.

Check the cables for evidence of corrosion in the form of a white or green powder around the battery ter-

minal end. If there is any doubt about the cleanliness of the battery terminals, pull the trouble codes by placing a jumper between terminals A and B of the ALCL connector, and monitoring the Check Engine light. Pulling the codes at this point will ensure that the information stored in the ECM's diagnostic memory will be retained for use later in the troubleshooting procedure. These codes will become the only way of determining any intermittent defects that may require attention after the hard faults have been diagnosed.

After writing down all of the fault codes, turn off all electrical loads on the car that can be turned off. Disconnecting the battery terminal with a load on the battery increases the possibility of damaging the ECM due to an induced voltage spike. Disconnect the terminals and clean or replace them as necessary.

Vacuum Hoses

Several of the sensors and actuators require either manifold or ported vacuum in order to function properly. Inspect all the vacuum hoses for evidence of swelling, cracks or leaking. Check for a snug fit at all connections and proper routing. Under the hood of most North American-built cars there is a vacuum-hose routing diagram. This diagram is often part of the EPA sticker. Compare the routing of the vacuum hoses to this chart.

Wiring Harness and Routing

Take a thorough look at the condition of the main engine wiring harness as well as the computer engine control harness. Look for broken or frayed wires, loose connections, damaged insulation and wires pinched under brackets or other engine parts.

Unfortunately, there is not a routing diagram for the wiring harness. Ensure that all of the computer or sensor wiring is at least 2 inches from secondary ignition wiring, and that the wiring harness is not stretched.

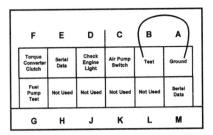

F	E	D	C	B	A
Torque Converter Clutch	Serial Data	Check Engine Light	Air Pump Switch	Test	Ground
Fuel Pump Test	Not Used	Not Used	Not Used	Not Used	Serial Data
G	H	J	K	L	M

The trouble codes can be pulled through the Check Engine light by placing a jumper wire between terminals A and B of the ALDL connector.

Fuel Leaks

The threat of fire from a fuel leak is obvious; however, there are other considerations. A fuel leak will cause a drastic reduction in fuel economy and may lower the fuel pressure enough to cause stumbling, hesitation or limited power at the top end.

Inspect around the injectors where they attach to the fuel rail, all of the fuel line connections at the filter and from the filter to the fuel rail. A place that is often overlooked is the fuel-pressure regulator. Remove the vacuum supply hose (TBI cars do not have one) from the fuel-pressure regulator. If there is fuel in the hose, replace the fuel-pressure regulator.

Exhaust Leaks

An exhaust system that is leaking out exhaust gases is also leaking in oxygen. Oxygen leaking into the exhaust system can cause the oxygen sensor to believe that the engine is running lean. When it reports this condition to the ECM it will respond by enriching the mixture. The end result will be a very rich running engine, poor fuel economy, fouled plugs and all of the symptoms that go with these conditions.

Scope Analysis

For the Saturday afternoon mechanic, the most difficult part of the testing procedure may be to get an accurate report. The car could be taken to the local garage that has one of those $25,000 wonder-machines. But to some extent, the accuracy of the results is highly dependent on

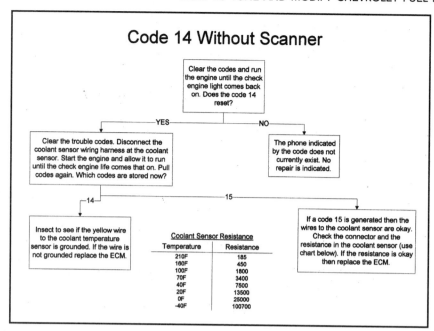

Code 14 Without Scanner

Clear the codes and run the engine until the check engine light comes back on. Does the code 14 reset?

YES → Clear the trouble codes. Disconnect the coolant sensor wiring harness at the coolant sensor. Start the engine and allow it to run until the check engine life comes that on. Pull codes again. Which codes are stored now?

NO → The phone indicated by the code does not currently exist. No repair is indicated.

14 → Insect to see if the yellow wire to the coolant temperature sensor is grounded. If the wire is not grounded replace the ECM.

15 → If a code 15 is generated then the wires to the coolant sensor are okay. Check the connector and the resistance in the coolant sensor (use chart below). If the resistance is okay then replace the ECM.

Coolant Sensor Resistance

Temperature	Resistance
210F	185
160F	450
100F	1800
70F	3400
40F	7500
20F	13500
0F	25000
-40F	100700

The manufacturer and many other sources (including this book) offer programmed diagnostic flow charts that offer the professional and the Saturday afternoon mechanic alike a logical progression of tests to reach an accurate conclusion.

Sensor Circuits

Reference Code	Circuit
S1	Switch to voltage
S2	Switch to pull low
S3	Variable resistance to pull low
S4	Variable resistance to push up
S5	Three-wire variable voltage
S6	DC frequency pulse generator
S7	AC rotational pulse generator
S8	DC rotational pulse generator
S9	Voltage generator

Actuator Circuits

Reference Code	Circuit
A10	Normally grounded
A11	Normally powered

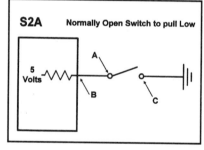

S1A   Normally Open Switch on Voltage

All of the sensor and actuator circuits in the fuel-injection system can be placed in 1 of 11 circuit types. The first of these circuits is the rarest: A circuit will supply 12 volts to the ECM microprocessor through terminal B when an event occurs. Two variations on this type of sensor will use either a normally-open or a normally-closed switch. If the normally open switch is used, voltage at B will remain low until the event occurs. The ECM monitors application of the air conditioner compressor clutch through this type of circuit.

S2A   Normally Open Switch to pull Low

A more common circuit consists of a function-controlled switch which when closed pulls an ECM-monitored reference voltage (either 5 or 12 volts) to ground (0 volts). This type of circuit is used to monitor a variety of functions in the fuel injection system such as EGR vacuum and transmission gear applications. Two variations on this circuit involve the use of a normally open switch which pulls the reference voltage low when the event being monitored occurs, and the normally closed switch which allows the voltage to rise when the event occurs. For the purposes of practical troubleshooting, we do not have to be concerned with whether the switch is normally open or normally closed, only with whether or not a change in voltage accompanies the event.

the operator of the equipment. As an option, refer to the tune-up chapter of this book.

Code Pulling

To pull the trouble codes, insert a jumper between terminals A and B of the ALCL connector. Monitor the Check Engine light for the flashing of the two-digit codes.

Fuel Pressure Check

The fact that checking fuel pressure is mentioned so often in this chapter on troubleshooting the Chevy fuel injection systems is no accident. Incorrect fuel pressure can cause a wide range of driveability symptoms.

Flow Chart Usage

All of the trouble codes and data fields we have looked at fall into 1 of 11 categories of circuits. What follows is a series of flow charts and diagrams that can be used along with the trouble code descriptions in chapter 8. A two-digit reference code such as S3 is found next to the trouble codes. Follow the flow chart that corresponds to the two-digit reference number. These flow charts have been designed for simplicity and convenience and will meet the needs of virtually all hard code causes. If by chance following these simplified charts does not lead to the source of the fault, consult the factory driveability book for the trouble code.

Sensor Circuits

There are nine types of sensor circuits used in Chevrolet fuel injection systems. Here are some simplified troubleshooting techniques. Use the following test procedure only when a code indicates a fault in one of the following circuit types.

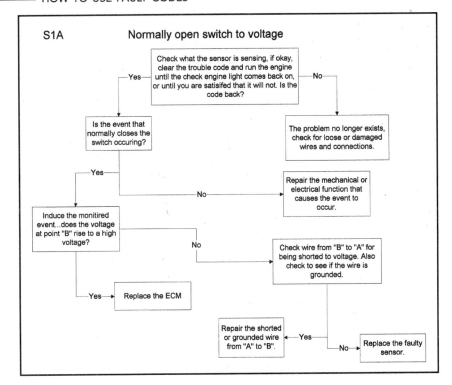

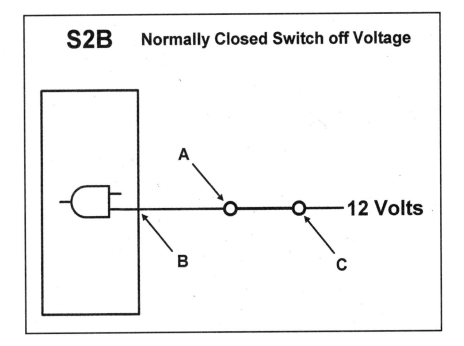

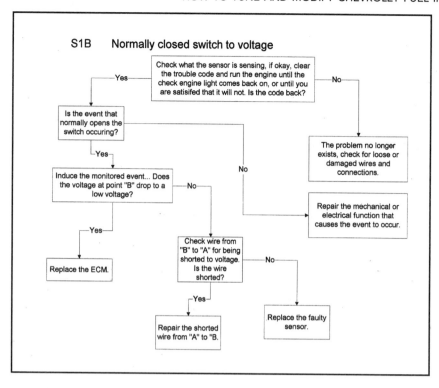

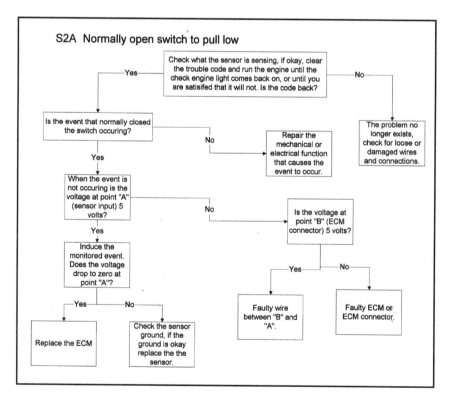

Sensor Circuits Reference
Sensor Circuit
Code Circuit
S1   Switch to voltage
S2   Switch to pull low
S3   Variable resistance to pull low
S4   Variable resistance to push up
S5   Three-wire variable voltage
S6   DC frequency pulse generator
S7   AC rotational pulse generator
S8   DC rotational pulse generator
S9   Voltage generator

Actuator Circuits Reference
Code Circuit
A10   Normally-grounded
A11   Normally-powered

S1 Sensor Circuit

To troubleshoot the switch-to-voltage circuit, connect a voltmeter to the terminal at the ECM. The voltage should read 0 if the monitored event is not occurring and a voltage if it is. Induce a change of state in the monitored device and a change in voltage should be observed. If the voltage remains the same, disconnect the monitored device and check resistance across the sensing switch.

The resistance should change from infinity to 0 ohms or 0 to infinity as the sensing switch is actuated. If this change does not occur, replace the sensor. If the change does occur, check the wire to the voltage source and the wire to the ECM for opens, shorts or grounds.

S2 Sensor Circuit

The switch-to-pull low circuit is tested by connecting a voltmeter to the input terminal of the ECM. As the monitored event occurs the voltage should go from 5 to 0 volts or from 0 to 5 volts. If the voltage remains at 5 volts, test the sensing switch with an ohmmeter. If the switch is good, look for an open in the wire between the ECM and the sensor or the sensor and ground. If the voltage remains at 0, disconnect the sensing switch. If the voltage rises to 5 volts, replace the sensor. If the voltage does not rise, inspect the wire from the ECM to the sensor for a short to ground. If the wire is good, replace the ECM.

S3 and S4 Sensor Circuits

The variable resistance-to-pull and the variable resistance-to-push up circuits should read between 0 and 5 volts if the circuit is functioning properly. If the voltage at the ECM is 5 volts, there is an open in the circuit. Check the continuity of the wire between the ECM and the sensor, as well as the wire between the sensor and ground. If the wire is good, replace the sensor. If there are 0 volts, check for a grounded wire between the ECM and the sensor. If this wire is not grounded and the sensor has more than 0 ohms of resistance, replace the ECM.

S5 Sensor Circuit

The three-wire, variable-voltage circuit will have one wire with reference 5 volts, a second wire with a constant ground (0 volts) and a third wire with a voltage greater than 0 but less than 5 volts. This third wire is the signal wire.

Connect a voltmeter to the ECM end of the third wire. If the voltage is between 0 and 5 and changes over a wide range as the condition being monitored changes, the circuit is good.

The Type 3 and Type 4 sensor circuits are virtually identical, with the Type 3 being the most common. When the monitored event occurs, resistance in the sensor decreases, causing a reference voltage to decrease. This is the type of circuit that is used to measure temperatures such as coolant and manifold air. Operating a little backwards from the usual way of thinking about circuit problems, this one—being a grounding circuit—will cause the voltage at B (the voltage seen by the ECM) to be high when the circuit is open. The Type 4 sensor circuit is the same, except the voltage begins low and as the monitored event occurs the voltage is pushed up.

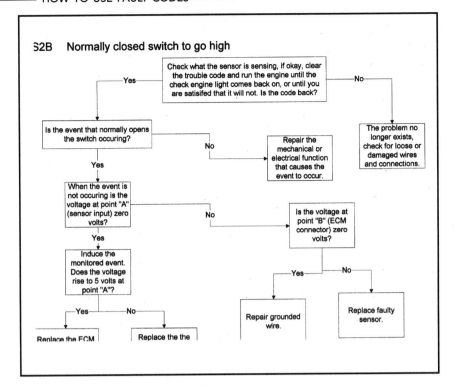

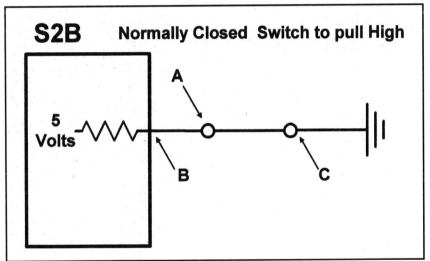

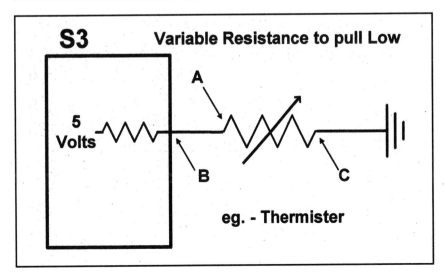

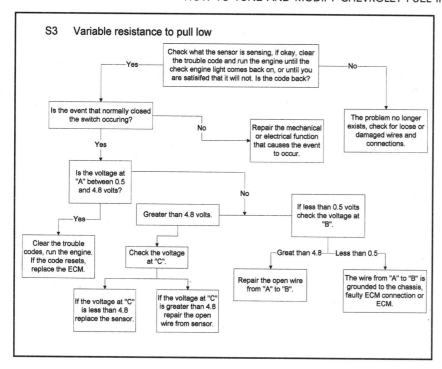

S3   Variable resistance to pull low

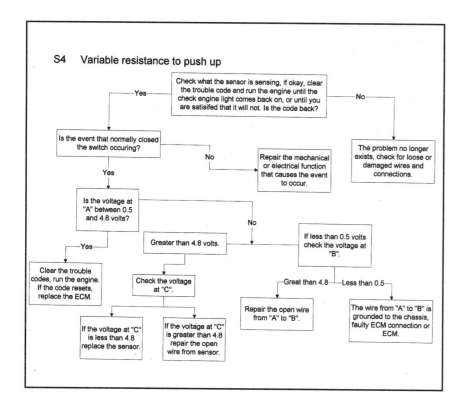

S4   Variable resistance to push up

If the voltage on the third wire is 0 or 5 at the ECM, check for an open circuit in the 5-volt reference or the signal wire. If these wires are okay, check the voltage at the ECM end of the 5-volt reference wire. If there is not 5 volts at the ECM, replace it. If there is, replace the sensor.

If the voltage on the third wire is less than 5, but greater than 0, and does not change significantly as the condition being monitored changes, inspect the sensor ground wire. If the sensor ground wire is not open, replace the sensor.

S6 Sensor Circuit

The DC frequency pulse generator can be tested with a high-impedance tachometer and voltmeter. The power supply might come from the ECM at 5 volts or from switched ignition at 12 volts. Begin by testing the power supply voltage and the sensor ground wire. If these portions of the circuit are good, connect a tachometer to the sensor output at the ECM. If there is a signal at the ECM, replace the ECM. If there is no pulse indicated on the tachometer, inspect the signal wire from the sensor to the ECM for opens, short or grounds. If the signal wire is good, replace the sensor.

S7 Sensor Circuit

Disconnect the rotational pulse generator (pickup coil) from the wiring harness. Connect an AC voltmeter to the output leads of the pickup and crank the engine. At cranking speeds, the AC volt-meter should read at least 0.5 volts AC. If the output is less than 0.5 volts, replace the pickup. If the cranking speed is greater than 200 rpm but the output is greater than 0.5, inspect the wiring between the pickup and the ignition module for opens, shorts, or grounds.

S8 Sensor Circuit

The DC rotational generator used by Chevrolet is called the Hall Effects sensor. The Hall Effects unit has three wires: The first wire carries the sensor power supply; the second is a ground; and the third is the signal.

Connect a tachometer to the signal wire and crank the engine. The tachometer should read something other than zero. If it does not, verify power to the sensor and verify the ground. If good, inspect the signal wire for opens, shorts and grounds. If the wire is good, replace the sensor.

Other Sensor Circuits

For the following codes refer only to the specific code flow chart: S9 indicates the Voltage Generator (oxygen sensor).

Actuators

A1 indicates a Normally-grounded actuator; and A2 indicates a Normally-powered actuator.

You Can Do It!

Whatever type of circuit the source of the problem might be in, the thing to keep in mind is that no circuit is difficult to troubleshoot. Think of it this way: The bulk of the sensor and actuator circuits have only two or three wires. Most people who consider themselves automotive enthusiasts or do-it-yourselfers can figure out why a high beam headlight bulb does not illuminate. These bulbs have only two terminals. One carries power; the other provides a ground. All of the two-wire sensors—with the exception of the rotational sensors—are no more complicated. A temperature-sensor circuit, for instance, can be diagnosed by disconnecting the sensor and connecting a voltmeter across the two wires. If it were a headlight, you would expect the voltmeter to read 12 volts; in the temperature sensing circuit, you would expect the reading to be 5 volts. If the voltage is correct, then the 5-volt reference wire is good and the ground wire is good. If not, then use the block for a ground to check the 5-volt wire. If there is 5 volts, then repair the ground wire; if there is not, repair the 5-volt wire. If the initial reading was 5 volts, then check the connector. If it is good, replace the sensor.

The three-wire, non-rotational sensors are only slightly more complex. When checking the wiring harness, treat the extra wire as though it were an extra ground. Check the voltage between the 5-volt reference wire and ground. This should read 5 volts. Check the voltage between the 5-volt reference and the third, or signal, wire. This should read slightly less than 5 volts. If this all tests good, check the connector, then replace the sensor. This is really no more complex than troubleshooting a high-beam headlight.

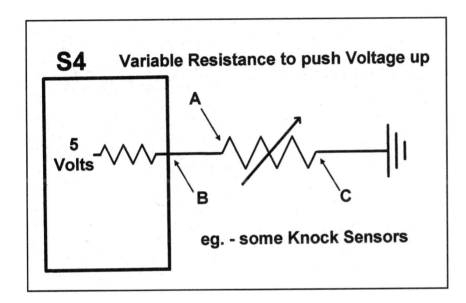

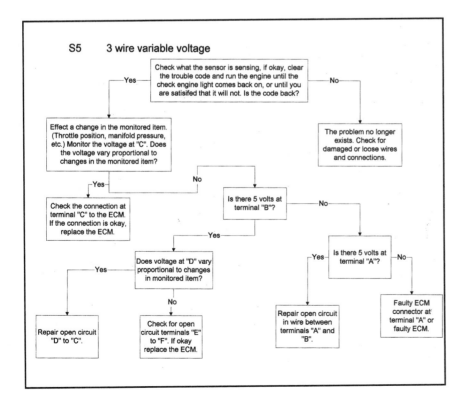

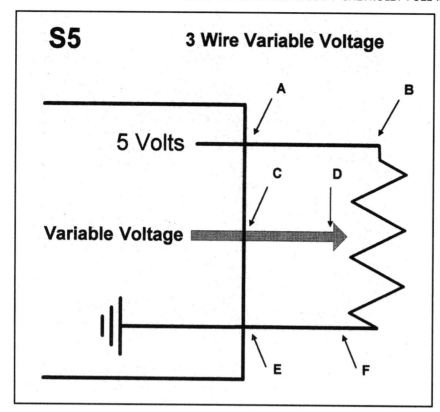

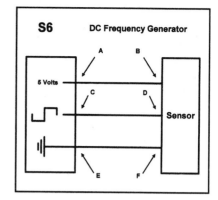

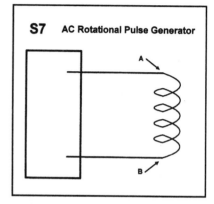

Type S5 is a three-wire sensor with a variable voltage signal to the ECM. The wire that runs between terminals A and B should have a constant 5 volts. The wire between terminals E and F should have a continuous ground (0 volts), while the wire between C and D will vary in voltage as the event being monitored varies. Included in the Type 5 sensor circuit are the TPS, the 3.1-liter EGR valve position sensor, the MAP sensor and the Bosch-style MAF used on TPI applications.

The AC rotational pulse generator, or pickup coil, is used to detect engine rpm on both distributor-type systems and the Direct Ignition System (DIS).

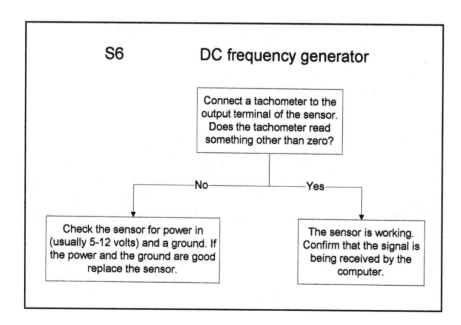

The Delco MAF produces a variable frequency DC pulse when sensing airflow. Check this frequency with a digital tachometer.

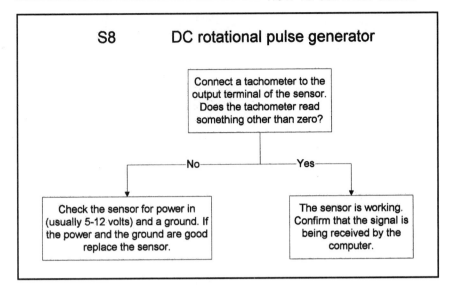

S8 — DC rotational pulse generator

Connect a tachometer to the output terminal of the sensor. Does the tachometer read something other than zero?

No → Check the sensor for power in (usually 5-12 volts) and a ground. If the power and the ground are good replace the sensor.

Yes → The sensor is working. Confirm that the signal is being received by the computer.

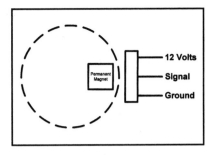

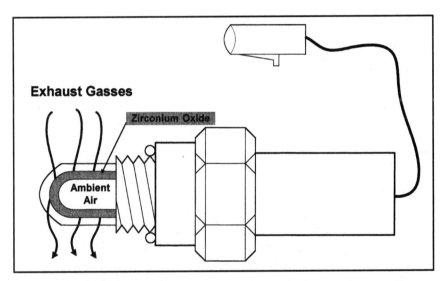

The oxygen sensor is a Type 9 sensor circuit. It is unique in that it creates its own voltage. The voltage output of the oxygen sensor ranges from 100 to 900 millivolts.

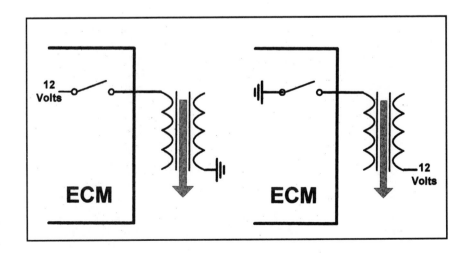

There are only two types of actuator circuits. The first of these would be circuit 10 which is a normally-grounded circuit. This is the least common of the two since it is difficult or impossible for the ECM to accurately monitor the functioning of the circuit. The second type, is the normally-powered actuator circuit. In this type of circuit the actuator has switched ignition voltage applied to it while the ECM makes and breaks the ground. This is the type of actuator circuit that Chevrolet uses most often.

C O N T E N T S

**TROUBLESHOOTING
WITH THE
VOLTMETER:**   **118**

**TROUBLESHOOTING
WITH THE
SCANNER**   **139**

CHAPTER TEN

TROUBLESHOOTING FAULT CODES

Fault or trouble codes are generated by the ECM whenever the computer detects a defect in one of its monitored circuits. We are going to describe the troubleshooting process for each of the codes. Wherever appropriate, variations in the testing procedures will be pointed out. Although an extraordinary attempt has been made to be complete, keep in mind that many variations exist and all may not be included for each code.

This code analysis chapter will be divided into two sections. The first section will outline the use of a voltmeter for this purpose, while the second will look at the use of the diagnostic scanner. As you work your way through either set of the trouble code test procedures, it is essential to keep in mind that it must be followed step by step, test by test, in the sequence described.

Troubleshooting with the Voltmeter:
Code 13

Oxygen sensor circuit testing should begin with checking the ECM's ability to go into closed loop. Start the engine and allow it to warm up by running at 2,000 rpm for 2 minutes. Allowing the engine to return to an idle, place a jumper wire

Since the voltages displayed on the scanner are the voltages that actually reach the ECM, if you have no access to a scanner the second best way to perform a scanner-based diagnosis is to connect the voltmeter as close to the ECM as possible. Here, the oxygen sensor voltage is being monitored.

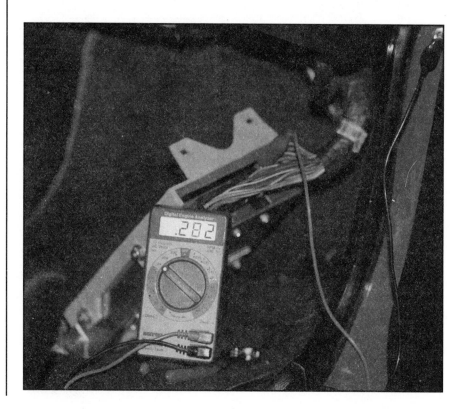

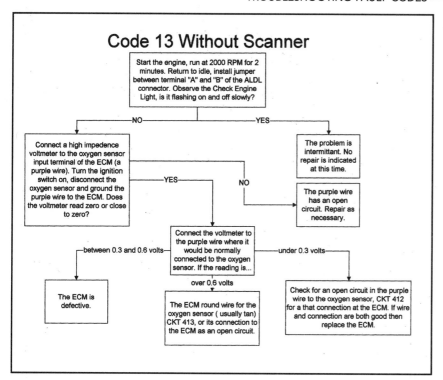

Code 13 Without Scanner

Start the engine, run at 2000 RPM for 2 minutes. Return to idle, install jumper between terminal "A" and "B" of the ALDL connector. Observe the Check Engine Light, is it flashing on and off slowly?

— NO — — YES —

NO branch: Connect a high impedence voltmeter to the oxygen sensor input terminal of the ECM (a purple wire). Turn the ignition switch on, disconnect the oxygen sensor and ground the purple wire to the ECM. Does the voltmeter read zero or close to zero?

YES branch: The problem is intermittant. No repair is indicated at this time.

— YES — — NO —

The purple wire has an open circuit. Repair as necessary.

Connect the voltmeter to the purple wire where it would be normally connected to the oxygen sensor. If the reading is...

— between 0.3 and 0.6 volts — — under 0.3 volts —

over 0.6 volts

The ECM is defective.

The ECM round wire for the oxygen sensor (usually tan) CKT 413, or its connection to the ECM as an open circuit.

Check for an open circuit in the purple wire to the oxygen sensor, CKT 412 for a that connection at the ECM. If wire and connection are both good then replace the ECM.

between terminals A and B of the ALDL connector and observe the Check Engine light. If the light is flashing on and off rather slowly, about every other second, then ECM has gone into closed loop and the problem that generated the code is intermittent. No repair is indicated at this time. If, however, the Check Engine light is flashing on and off about two times per second, then the ECM has not entered closed loop and the problem still exists.

Connect a high-impedance (10 megohm input impedance or more) voltmeter to the oxygen sensor input terminal at the ECM. Turn the ignition switch on. Disconnect the oxygen sensor and ground the purple wire that runs from the oxygen sensor to the ECM. Does the voltmeter read close to 0 or 0 volts? If it does not, then the purple wire has an open

If the "Check Engine" or "Service Engine Soon" light is on while the engine is running, it means that at least one of the trouble codes set in memory is "hard." A hard code is a fault that is happening at that very moment, and should be easy to trace using the flow charts.

The coolant temperature sensor is usually located near the thermostat housing or near the upper radiator hose. Shown here is the location on a 1984 Citation.

circuit and should be repaired. If the voltmeter does read 0 volts, this proves that the wire is good.

Now connect the purple wire to the voltmeter connector that would normally be connected to the oxygen sensor. With the ignition switch still on, make these determinations:

- If the voltmeter reads between 0.3 and 0.6 volts, then the ECM is defective.
- If the voltmeter reads over 0.6 volts, the ECM ground wire for the oxygen sensor (CKT 413, usually tan) or its connection to the ECM has an open circuit. If inspection of the wire and connection indicates no problem, replace the ECM.
- If the voltmeter reads below 0.3 volts, then there is an open in the purple wire (CKT 412). To check

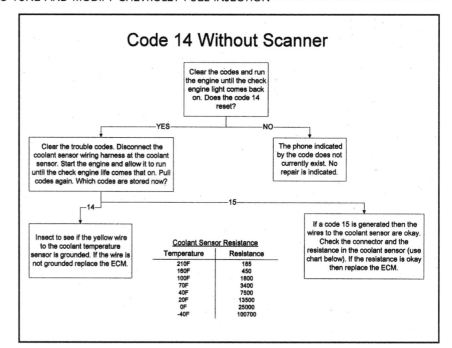

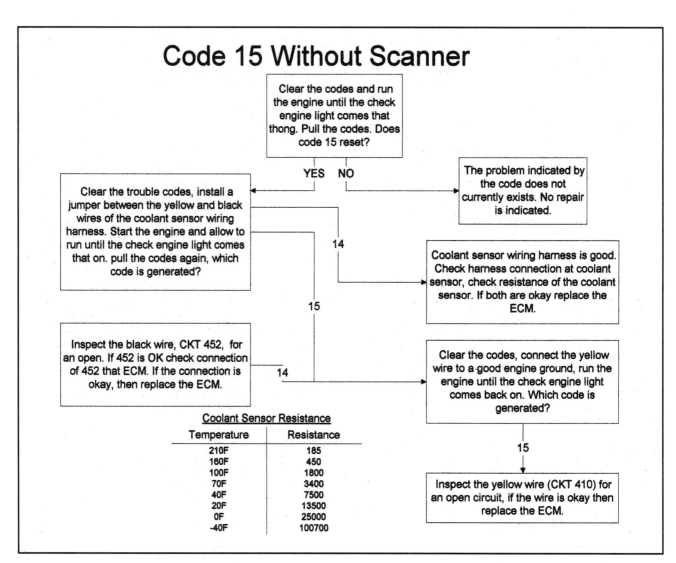

for a faulty connection for the purple wire at the ECM, inspect the wire and connection. If they are okay, replace the ECM.

Code 14

Code 14 indicates that the computer has seen or is seeing a low voltage on the coolant sensor wire, CKT 410 (the yellow wire).

Clear the codes and run the engine without driving the car until the Check Engine light comes back on, or until you are satisfied that it will not. Does Code 14 reset? If not, the problem is intermittent and no repair is indicated at this time.

Clear the trouble codes. Disconnect the coolant sensor wiring harness at the coolant sensor. Start the engine and allow to run until the

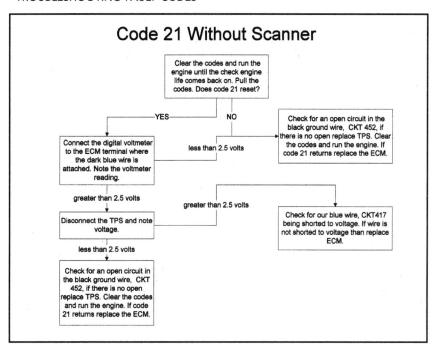

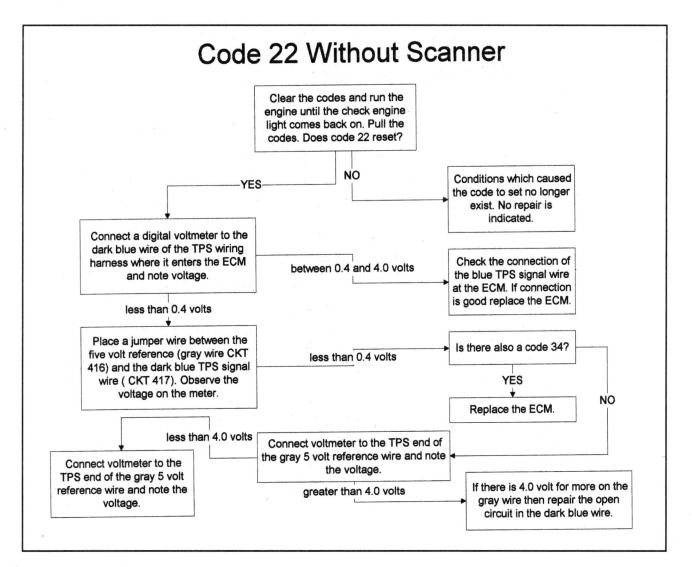

Check Engine light comes back on. Pull the codes.

If Code 14 returns, the yellow wire to the coolant sensor (CKT 410) is grounded.

If a Code 15 comes up, the wiring harness is good. Check the coolant sensor connector and coolant sensor resistance (see coolant sensor resistance chart). If both check out, then replace the ECM.

Code 15

Code 15 indicates that the ECM is seeing a high voltage on the wire to the Coolant Temperature Sensor, CKT 410 (the yellow wire).

Clear the codes and run the engine until the Check Engine light comes back on or until you are satisfied that it will not. Does Code 15 return? If not, then the problem is intermittent and no repair is indicated at this time.

If the Code 15 returns, clear the trouble codes, disconnect the coolant temperature sensor wiring harness from the coolant sensor and install a jumper between the two wires of the coolant sensor harness, CKT 410 and 452 (usually a yellow wire and a black wire). Start the engine and allow it to run until the Check Engine light comes back on. Then pull the codes. Is there a Code 15 or 14?

If there is a Code 14, the coolant sensor wiring harness is okay. Check the connection to the coolant sensor. Check the resistance in the coolant sensor. If both are okay, then replace the ECM.

If there is a Code 15, the wiring harness has an open circuit. Clear the codes, connect a jumper wire from the yellow wire of the coolant sensor harness to engine block ground, start the engine and allow it to run until the Check Engine light comes back on, then pull the codes. If Code 14 comes up, inspect the black wire (CKT 452) for an open. If Code 15 comes up, inspect the yellow wire (CKT 410) for an open. If either 14 or 15 were generated, but neither wire has an open circuit, replace the ECM.

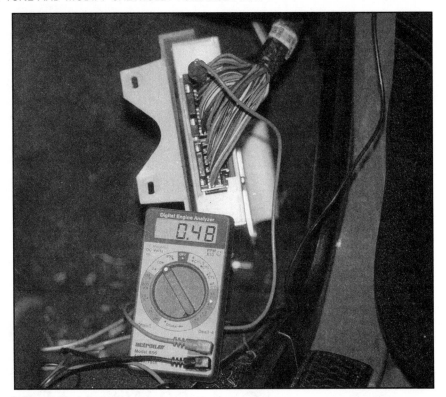

Checking the TPS input voltage at the ECM. If the voltage is equal to the correct closed-throttle voltage, then the entire TPS circuit is good. An open or grounded 5-volt reference wire would result in 0 volts at the ECM input, as would an open or grounded signal wire. A voltage greater than 1 but less than 4 that changes very little as the throttle is opened and closed may indicate an open in the TPS ground wire.

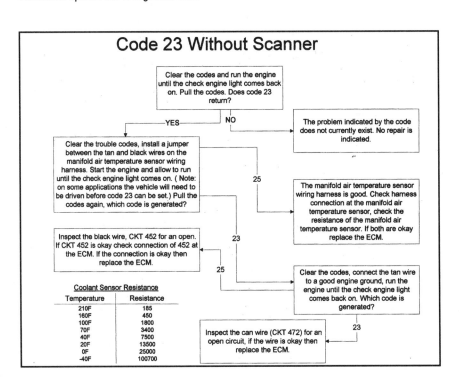

Code 23 Without Scanner

Clear the codes and run the engine until the check engine light comes back on. Pull the codes. Does code 23 return?

YES — NO

The problem indicated by the code does not currently exist. No repair is indicated.

Clear the trouble codes, install a jumper between the tan and black wires on the manifold air temperature sensor wiring harness. Start the engine and allow to run until the check engine light comes on. (Note: on some applications the vehicle will need to be driven before code 23 can be set.) Pull the codes again, which code is generated?

25

The manifold air temperature sensor wiring harness is good. Check harness connection at the manifold air temperature sensor, check the resistance of the manifold air temperature sensor. If both are okay replace the ECM.

Inspect the black wire, CKT 452 for an open. If CKT 452 is okay check connection of 452 at the ECM. If the connection is okay then replace the ECM.

23

25

Clear the codes, connect the tan wire to a good engine ground, run the engine until the check engine light comes back on. Which code is generated?

Coolant Sensor Resistance	
Temperature	Resistance
210F	185
160F	450
100F	1800
70F	3400
40F	7500
20F	13500
0F	25000
-40F	100700

Inspect the can wire (CKT 472) for an open circuit, if the wire is okay then replace the ECM.

23

Code 21

Code 21 will set when the ECM sees a high-signal voltage on the TPS signal wire, usually dark blue (CKT 417).

Begin the troubleshooting of a Code 21 by clearing the codes and running the engine until the Check Engine light either comes on, or you are sure that it will not. If the light does not come back on, then the condition that set the code no longer exists and no repair is indicated.

If Code 21 recurs, connect a voltmeter to the TPS signal wire where it enters the ECM. Turn the ignition switch to the "On" position and note the voltmeter reading. If the reading is less than 2.5 volts, the condition is intermittent and does not exist at the present time. If the reading is greater than 2.5 volts, disconnect the TPS wiring harness and note the voltmeter.

If the TPS voltage drops below 2.5 volts when the harness is disconnected, check for an open circuit in the black ground wire, CKT 452. If there is no open, then replace the TPS. Clear the codes and run the engine. If Code 21 returns, replace the ECM.

If the TPS voltage remains above 2.5 volts when the TPS wiring harness is disconnected, check for a short to voltage on the TPS signal wire, CKT 417 (dark blue wire). If the dark blue wire is not shorted to voltage, then replace the ECM.

Code 22

Code 22 sets when the voltage on the dark blue wire (CKT 417) of the TPS wiring harness is seen by the ECM to have an exceptionally low voltage.

Clear the codes and run the engine until the Check Engine light comes back on. If the light does not come back on, the problem is intermittent.

If the Code 22 comes back, connect a digital voltmeter to the TPS signal wire (dark blue CKT 417) where it enters the ECM. With the key on and engine off, observe the

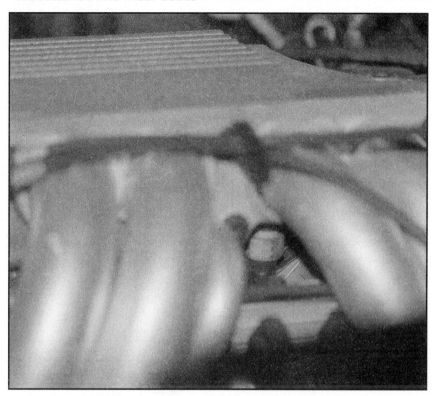

On the 5.7-liter Tuned Port Injection Corvette engine, the testing of the MAT is awkward. Working through the Code 23 and 25 flow charts can be frustrating. The same is true of the 5.0-liter TPI applications.

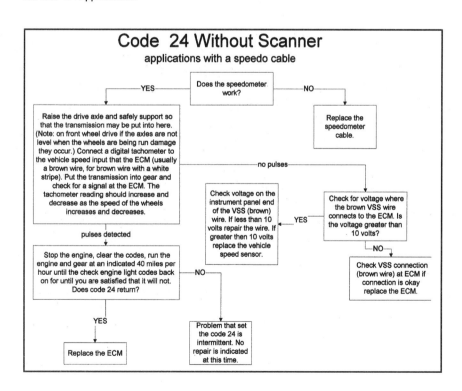

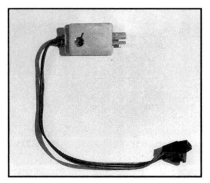

The vehicle speed sensor (VSS) used on the cars equipped with a speedometer cable consists of a light emitting diode/photo diode pair (the black end) which mounts in the speedometer head, and a buffer (light-colored end) which sends a square wave to the ECM directly proportional to the speed of the car.

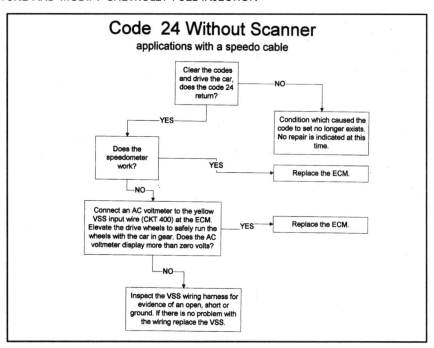

Code 24 Without Scanner
applications with a speedo cable

Clear the codes and drive the car, does the code 24 return?

— NO → Condition which caused the code to set no longer exists. No repair is indicated at this time.

— YES → Does the speedometer work?

— YES → Replace the ECM.

— NO → Connect an AC voltmeter to the yellow VSS input wire (CKT 400) at the ECM. Elevate the drive wheels to safely run the wheels with the car in gear. Does the AC voltmeter display more than zero volts?

— YES → Replace the ECM.

— NO → Inspect the VSS wiring harness for evidence of an open, short or ground. If there is no problem with the wiring replace the VSS.

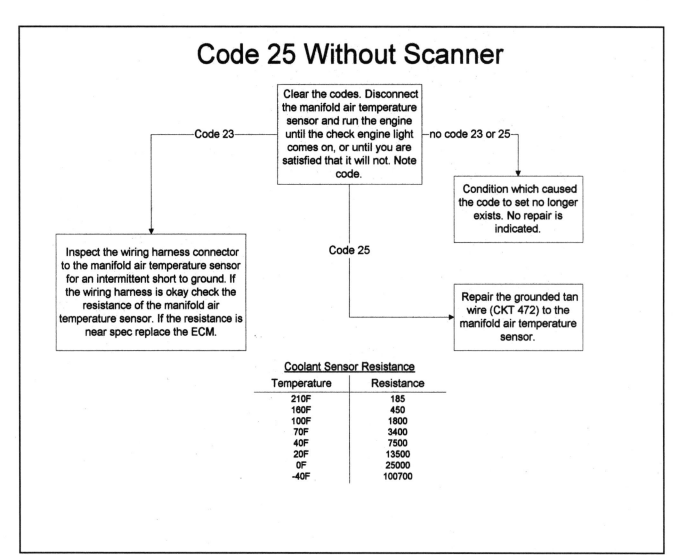

Code 25 Without Scanner

Clear the codes. Disconnect the manifold air temperature sensor and run the engine until the check engine light comes on, or until you are satisfied that it will not. Note code.

— Code 23 → Inspect the wiring harness connector to the manifold air temperature sensor for an intermittent short to ground. If the wiring harness is okay check the resistance of the manifold air temperature sensor. If the resistance is near spec replace the ECM.

— no code 23 or 25 → Condition which caused the code to set no longer exists. No repair is indicated.

— Code 25 → Repair the grounded tan wire (CKT 472) to the manifold air temperature sensor.

Coolant Sensor Resistance

Temperature	Resistance
210F	185
160F	450
100F	1800
70F	3400
40F	7500
20F	13500
0F	25000
-40F	100700

voltage reading. If the reading is between 0.4 and 4.0 volts, check the connection of the dark blue wire at the ECM. If the connection is okay, replace the ECM.

If the voltage is less than 0.4 volts, disconnect the TPS harness and place a jumper wire between the 5-volt reference to the TPS (gray wire CKT 416) and note the voltage.

If the voltmeter (still connected to the dark blue wire at the ECM) reads less than 0.4 and there is also a Code 34, replace the ECM. If it reads less than 0.4 and there is no Code 34, connect the voltmeter to the TPS harness end of the gray 5-volt reference wire. If there is less than 0.4 volts on this wire, repair the open cir-

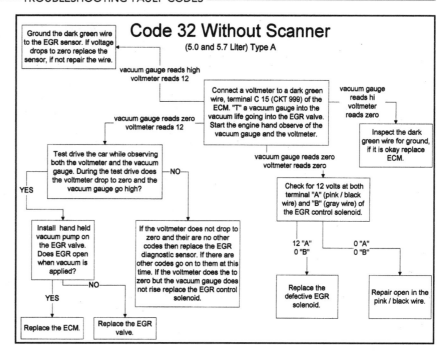

Code 32 Without Scanner
(5.0 and 5.7 Liter) Type A

Ground the dark green wire to the EGR sensor. If voltage drops to zero replace the sensor, if not repair the wire.

vacuum gauge reads high voltmeter reads 12

Connect a voltmeter to a dark green wire, terminal C 15 (CKT 999) of the ECM. "T" a vacuum gauge into the vacuum life going into the EGR valve. Start the engine hand observe of the vacuum gauge and the voltmeter.

vacuum gauge reads hi voltmeter reads zero

vacuum gauge reads zero voltmeter reads 12

Inspect the dark green wire for ground, if it is okay replace ECM.

Test drive the car while observing both the voltmeter and the vacuum gauge. During the test drive does the voltmeter drop to zero and the vacuum gauge go high?

vacuum gauge reads zero voltmeter reads zero

Check for 12 volts at both terminal "A" (pink / black wire) and "B" (gray wire) of the EGR control solenoid.

YES — NO

Install hand held vacuum pump on the EGR valve. Does EGR open when vacuum is applied?

If the voltmeter does not drop to zero and their are no other codes then replace the EGR diagnostic sensor. If there are other codes go on to them at this time. If the voltmeter does the to zero but the vacuum gauge does not rise replace the EGR control solenoid.

12 "A" 0 "B" / 0 "A" 0 "B"

Replace the defective EGR solenoid.

Repair open in the pink / black wire.

YES — NO

Replace the ECM.

Replace the EGR valve.

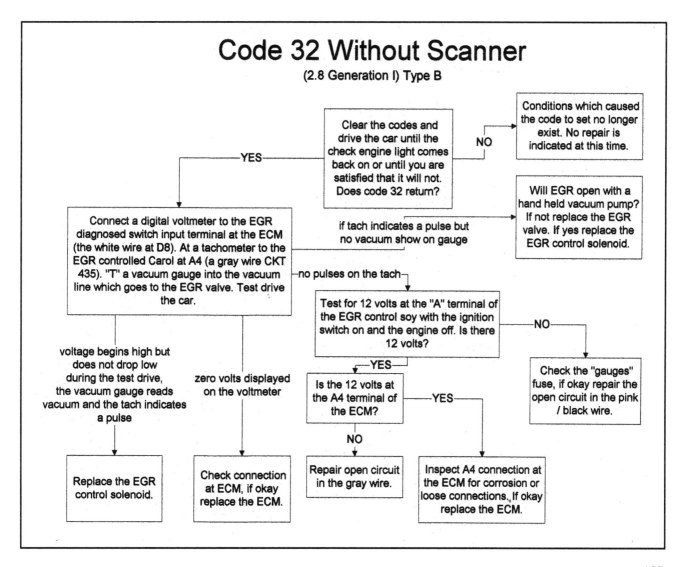

Code 32 Without Scanner
(2.8 Generation I) Type B

Clear the codes and drive the car until the check engine light comes back on or until you are satisfied that it will not. Does code 32 return?

NO

Conditions which caused the code to set no longer exist. No repair is indicated at this time.

YES

Will EGR open with a hand held vacuum pump? If not replace the EGR valve. If yes replace the EGR control solenoid.

Connect a digital voltmeter to the EGR diagnosed switch input terminal at the ECM (the white wire at D8). At a tachometer to the EGR controlled Carol at A4 (a gray wire CKT 435). "T" a vacuum gauge into the vacuum line which goes to the EGR valve. Test drive the car.

if tach indicates a pulse but no vacuum show on gauge

no pulses on the tach

Test for 12 volts at the "A" terminal of the EGR control soy with the ignition switch on and the engine off. Is there 12 volts?

NO

voltage begins high but does not drop low during the test drive, the vacuum gauge reads vacuum and the tach indicates a pulse

zero volts displayed on the voltmeter

YES

Is the 12 volts at the A4 terminal of the ECM?

YES

Check the "gauges" fuse, if okay repair the open circuit in the pink / black wire.

NO

Replace the EGR control solenoid.

Check connection at ECM, if okay replace the ECM.

Repair open circuit in the gray wire.

Inspect A4 connection at the ECM for corrosion or loose connections. If okay replace the ECM.

cuit in the gray wire or at the connection. If no open is found, then replace the ECM.

If there is 4.0 volts or more on the gray wire, repair the open circuit in the dark blue wire.

Code 23

Code 23 relates to a perceived high voltage in the Manifold Air Temperature terminal at the ECM.

Begin by clearing the codes and running the engine until the Check Engine light comes back on. Pull the codes again. If Code 23 did not

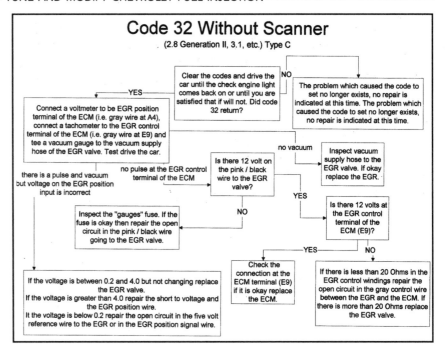

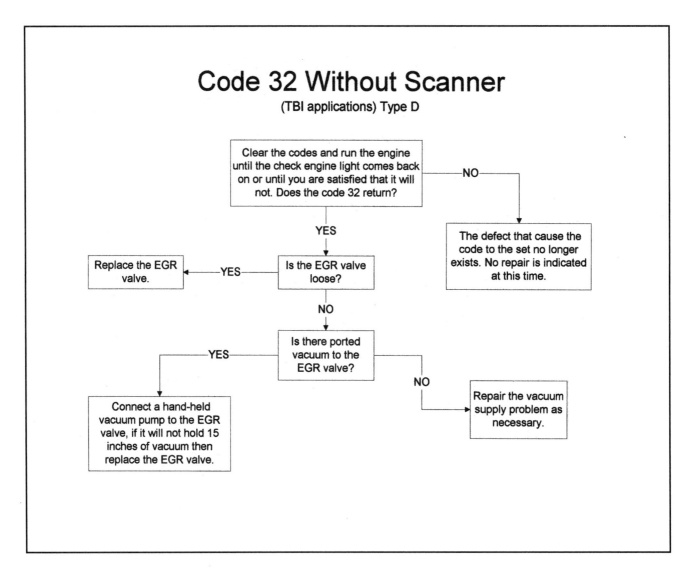

return, then the condition that caused Code 23 to set no longer exists and therefore no repair is indicated.

If the Code 23 returns, clear the trouble codes, disconnect the Manifold Air Temperature sensor wiring harness from the MAT sensor and install a jumper between the two wires of the MAT sensor harness. CKT 472 and 452 (usually a tan wire and a black wire). Start the engine and allow it to run until the Check Engine light comes back on. Pull the codes. Is there a Code 23 or 25?

If there is a Code 25, then the MAT sensor wiring harness is okay. Check the connection to the MAT sensor; then check the resistance in the MAT sensor. If both are okay,

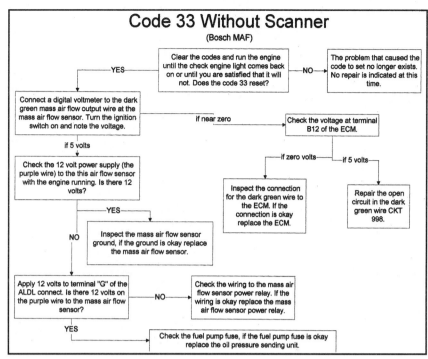

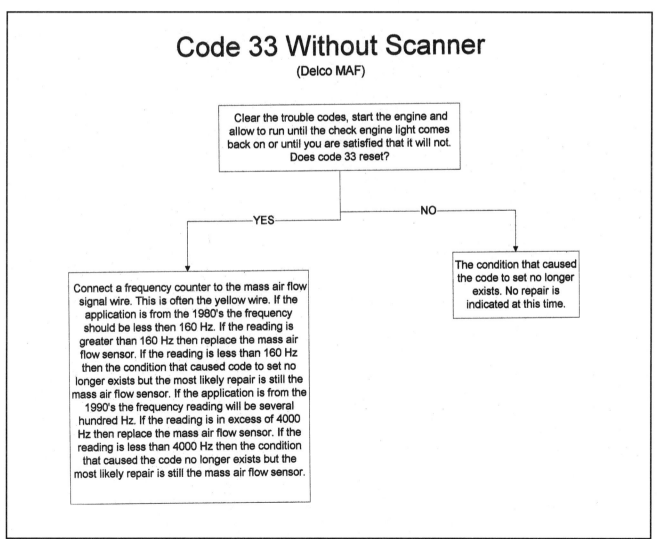

then replace the ECM.

If there is a Code 23, the wiring harness has an open circuit. Clear the codes. Connect a jumper wire from the tan wire of the MAT sensor harness to engine block ground. Start the engine and allow it to run until the Check Engine light comes back on. Pull the codes. If Code 25 comes up, inspect the black wire (CKT 452) for an open. If Code 23 comes up, inspect the tan wire (CKT 472) for an open. If either 23 or 25 were generated but neither wire has an open circuit, replace the ECM.

Code 24

The Vehicle Speed Sensor (VSS) is monitored by Code 24. There are two major categories of VSSs used by Chevrolet. The first,

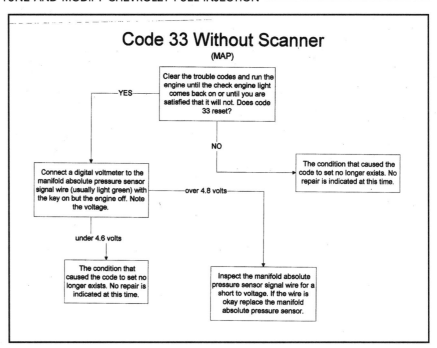

Code 33 Without Scanner (MAP)

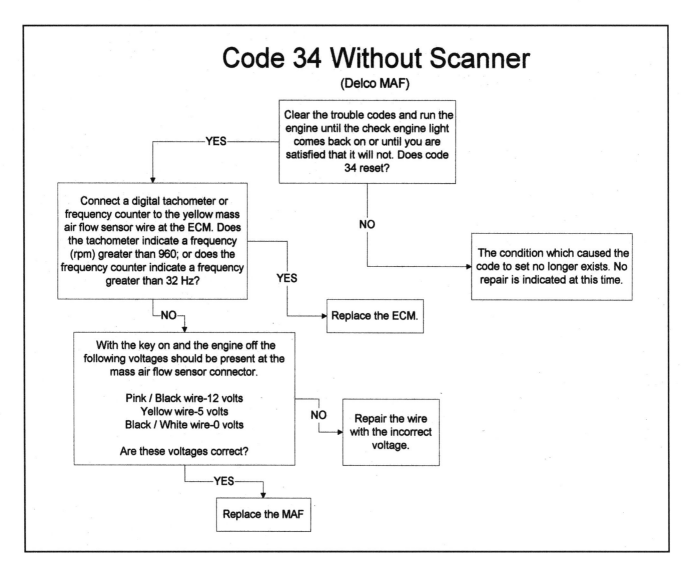

Code 34 Without Scanner (Delco MAF)

Type A, is used on applications that have a speedometer cable. This consists of an infrared light emitting diode (LED) and a photodiode. This diode pair mounts in the speedometer head where the light from the LED is reflected back to the photodiode by a notched rotating disc driven by the speedometer cable. The second type, Type B, is a magnetic pulse generator that feeds directly into the ECM. The ECM then supplies the speed signal to the speedometer.

Type A—Applications with Speedometer Cable

Disregard a Code 24 when it is set without the wheels turning.

Does the speedometer work? If the speedo does not work and there is a Code 24, replace the speedo cable.

Raise the drive wheels and support the drive axle so that the transmission may be safely put into gear to drive the wheels. (Note: Allowing the drive axles to hang while running the wheels may cause damage to driveline components. Support the drive axles so that they are in their normal horizontal position while running them.) Connect a digital tachometer to the vehicle speed sensor input wire at the ECM (brown wire CKT 437). Start the engine and put the car into gear. The tachometer should read something other than zero, and the reading should increase as the speed of the wheels increases.

If there are no pulses, check for voltage at the VSS terminal of the ECM. If the voltage on this wire (usually brown or brown and white, CKT 437) is greater than 10 volts at the ECM, check the voltage at the instrument panel end of the wire. If the voltage is less than 10 volts, repair the defective wire. If it is greater than 10 volts, replace the VSS.

If the voltage is less than 10 volts at the ECM, check the ECM connection. If the connection is okay, then replace the ECM.

If the brown wire has a pulse on it, clear the codes. With the drive axles still raised, start the engine, put

The printed circuitry of the Delco MAF produces a variable frequency square wave. Accurate troubleshooting of this sensor will require both a voltmeter and a digital tachometer.

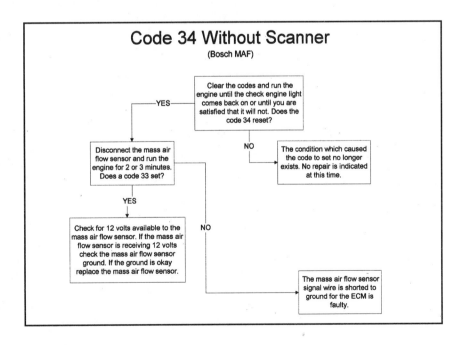

Code 34 Without Scanner
(Bosch MAF)

Disconnect the mass air flow sensor and run the engine for 2 or 3 minutes. Does a code 33 set?

YES → Clear the codes and run the engine until the check engine light comes back on or until you are satisfied that it will not. Does the code 34 reset?

NO → The condition which caused the code to set no longer exists. No repair is indicated at this time.

YES → Check for 12 volts available to the mass air flow sensor. If the mass air flow sensor is receiving 12 volts check the mass air flow sensor ground. If the ground is okay replace the mass air flow sensor.

NO → The mass air flow sensor signal wire is shorted to ground for the ECM is faulty.

the transmission in gear and run the wheels at an indicated 40 mph until the Check Engine light comes back on or until you are satisfied that it will not. If the Code 24 does not return, the problem is intermittent and no repair is indicated. If the Code 24 returns, replace the ECM.

Type B—Applications without Speedometer Cable

Does the speedometer work? Check the wire between the speedometer head and the ECM. If yes, then replace the ECM. If no, connect an AC voltmeter to the yellow wire VSS input (CKT 400) to the ECM. Safely elevate the drive wheels, put the transmission in gear and run the drive wheels. The AC voltmeter should display a voltage

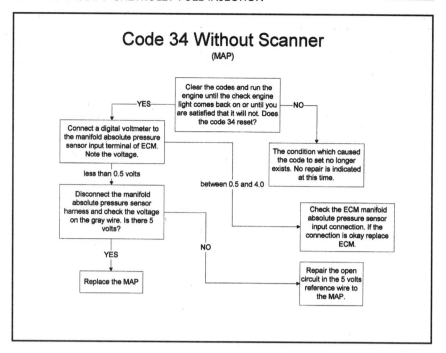

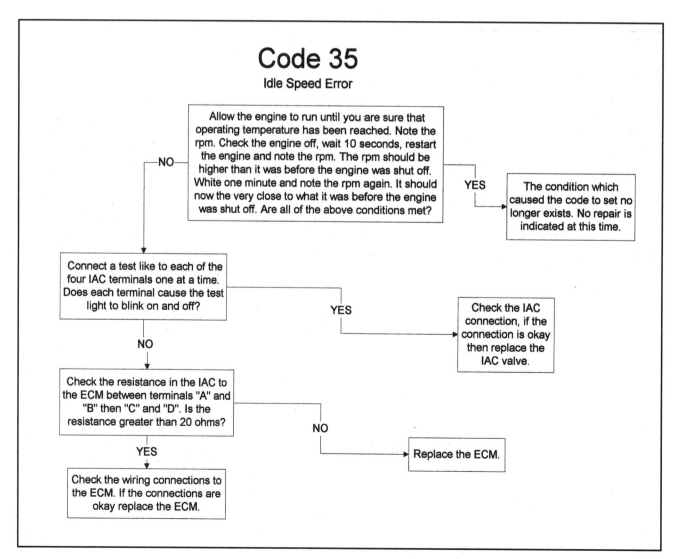

above 0. If it does not, inspect VSS wiring harness for an open, short, or ground. If the wiring harness is okay, replace the VSS.

Code 25

For a Code 25 to be set, the ECM must see close to 0 volts on the tan MAT sensor wire (CKT 472).

Clear the codes. Disconnect the harness at the MAT sensor. Start the engine and run until the Check Engine light comes back on. Pull the codes.

If a Code 23 is generated, inspect the wiring harness connector to the MAT for an intermittent short to ground. If the harness is okay, check the resistance in the MAT. If the resistance of the MAT is near specification, replace the ECM.

If a Code 25 is generated, repair the grounded tan wire to the MAT (CKT 472).

Code 26

Some applications are capable of generating a Code 26—an indication that there is a perceived failure in one of the quad-driver circuits. Since this diagnostic routine is extremely complex and easily prone to error, I recommend letting the dealer trace the source of this code.

Code 32

Code 32 relates to a perceived failure of the EGR system to function. On a 5.7-liter TPI system Code 32 means that the temperature of the sensor in the EGR tube has not increased, therefore the ECM does not know if the EGR valve opened when it was commanded to. The same is true on the 5.0-liter TPI engine.

The older 2.8-liter applications used a vacuum switch to determine if vacuum had been applied to the EGR valve. Later 2.8-liter applications (the Generation II), as well as the 3.0-liter use a potentiometer that not only monitors if the EGR is opening but also monitors the exact position of the valve.

We shall designate these three types of Code 32 sensors as Type A,

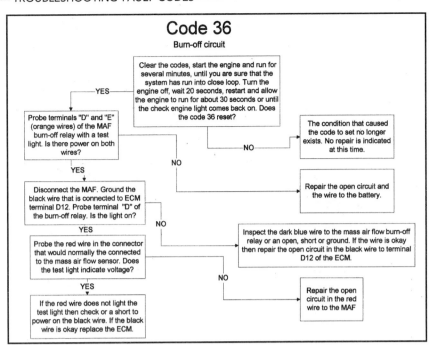

Code 36
Burn-off circuit

Clear the codes, start the engine and run for several minutes, until you are sure that the system has run into close loop. Turn the engine off, wait 20 seconds, restart and allow the engine to run for about 30 seconds or until the check engine light comes back on. Does the code 36 reset?

— YES —

Probe terminals "D" and "E" (orange wires) of the MAF burn-off relay with a test light. Is there power on both wires?

YES

Disconnect the MAF. Ground the black wire that is connected to ECM terminal D12. Probe terminal "D" of the burn-off relay. Is the light on?

YES

Probe the red wire in the connector that would normally the connected to the mass air flow sensor. Does the test light indicate voltage?

YES

If the red wire does not light the test light then check or a short to power on the black wire. If the black wire is okay replace the ECM.

NO — The condition that caused the code to set no longer exists. No repair is indicated at this time.

Repair the open circuit and the wire to the battery.

Inspect the dark blue wire to the mass air flow burn-off relay or an open, short or ground. If the wire is okay then repair the open circuit in the black wire to terminal D12 of the ECM.

Repair the open circuit in the red wire to the MAF

Code 41
without a scanner

Cylinder select error. If the calibration chip (MEMCAL, PROM) has been recently replaced then the code has been set due to improper programming of the chip. Check for proper seating of the chip and check for corrosion. If okay replace the calibration chip.

B, and C respectively. There is also a Type D that applies only to some TBI applications.

Type A—TPI 5.0- and 5.7-Liter Engines

Connect a digital voltmeter to the dark green wire at terminal C15 (CKT 999) of the ECM. Install a vacuum gauge into the vacuum line going to the EGR valve using a "T" adapter. Test drive the car. Observe both the vacuum gauge and the voltmeter. As you begin to drive the car, the voltmeter should indicate 12+ volts. When the EGR valve is commanded to open by the ECM, the vacuum gauge should indicate vacuum and the voltmeter voltage should drop.

If the vacuum gauge indicates vacuum but the voltmeter continues to read 12 volts, then the problem is in the sensor circuit. After the test drive, leave the voltmeter connected, disconnect the dark green sensor wire (CKT 999) at the sensor and ground it. If the voltmeter drops to 0, inspect the sensor connection. If the sensor connection is okay, replace the sensor. If the voltage remains at 12 volts, repair the open circuit in the dark green wire (CKT 999).

If the vacuum gauge did not indicate vacuum during the test drive, use the underhood vacuum routing diagram to identify and locate the EGR vacuum control solenoid. Connect the vacuum gauge to the inbound side of the solenoid and repeat the test drive. If the vacuum gauge still indicates no vacuum, repair the broken or mis-routed vacuum line. If vacuum is indicated, reconnect the vacuum hose to the EGR control solenoid and then install the vacuum gauge at the end of the hose going to the EGR valve. Repeat the test drive. If vacuum is read on the gauge, replace the EGR valve.

If no vacuum is read, connect the voltmeter to terminal A4 (gray wire, CKT 435) of the ECM. With key on and engine off there should be volts at this terminal. If there is not, check for voltage at the A (pink and black

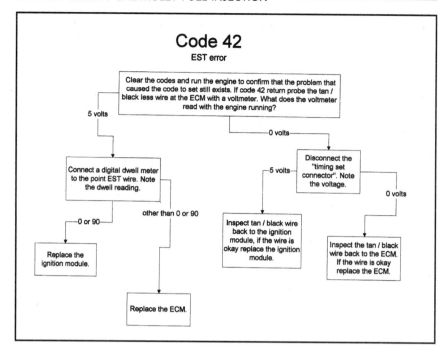

wire) and B (gray wire) terminals of the EGR control solenoid. If there is 12 volts on A but not B, replace the EGR solenoid. If there is no voltage on either terminal, inspect the gauges fuse. If the fuse is okay, repair open CKT 39 (pink and black wire).

Type B—2.8-Liter, Generation I

The Type B system uses a vacuum switch to determine if the EGR valve

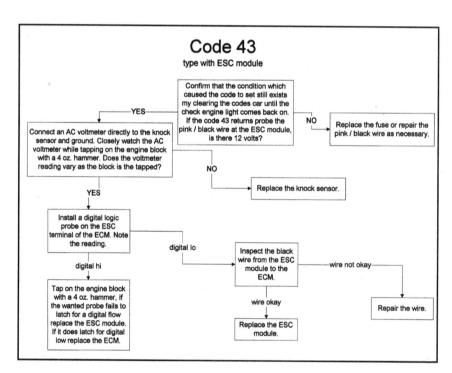

is receiving vacuum. This vacuum switch is mounted in a single unit along with the EGR control solenoid.

Connect a voltmeter to the EGR diagnostic switch input at terminal D8 (CKT 935, a white wire) of the ECM. Connect a digital tachometer to terminal A4 (CKT 435, a gray wire) and "T" a vacuum gauge into the vacuum line going to the EGR valve. Ensure that the engine is thoroughly warmed up, then test drive the car.

If a pulse is detected at A4, but no vacuum is detected, test the EGR with a hand-held vacuum pump. If the EGR valve opens with the hand-held pump, replace the EGR control solenoid.

If no pulse is detected at A4, check for a 12-volt power supply at the A terminal of the EGR control solenoid with the ignition switch on and engine off. If there is a 12-volt supply, then replace the EGR control solenoid. If there is not, check the gauges fuse. If that is working properly, repair the open circuit in the pink and black wire, CKT 39.

If the voltmeter does not have 12 volts on it at the beginning of the test drive, check the connection at ECM terminal D8. If the connection is okay, replace the ECM.

If the voltage begins high but does not drop low during the test drive, but the vacuum gauge does read vacuum and the tachometer does read a pulse, replace the EGR control solenoid.

Type C

The Type C EGR control system uses an integrated EGR valve-control solenoid-position sensor unit.

If a Code 32 is stored in the ECM, clear the codes and run the engine until the Check Engine light comes back on. If the code does not return, then the condition that caused the code to set no longer exists and no repair is indicated. If the Code 32 returns, connect a voltmeter to the EGR position sensor terminal of the ECM (CKT 474, a gray wire usually at ECM terminal A4). Connect a tachometer to terminal E9 (the gray wire, CKT 435). Connect a vacuum gauge to the vacuum supply hose to the EGR. Test drive the car.

If there is no pulse at the EGR control terminal of the ECM at any time during the test drive (CKT 435), check for voltage at the pink and black wire connected to the EGR valve. If there is 12 volts, check for voltage at the ECM terminal for the gray wire from the EGR to the EGR

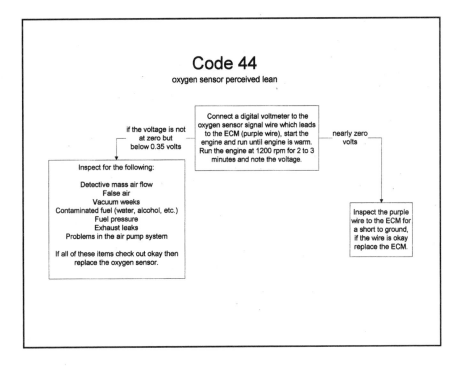

Code 43
integrated ESC circuit

Disconnect the knock sensor wiring connector, check the voltage on the wire going back to the ECM. If there is 5 volts on the wire replace the knock sensor. If there is not inspect the wire for an open, short or ground. If the wire is okay replace the ECM.

Code 44
oxygen sensor perceived lean

if the voltage is not at zero but below 0.35 volts

Connect a digital voltmeter to the oxygen sensor signal wire which leads to the ECM (purple wire), start the engine and run until engine is warm. Run the engine at 1200 rpm for 2 to 3 minutes and note the voltage.

nearly zero volts

Inspect for the following:

Detective mass air flow
False air
Vacuum weeks
Contaminated fuel (water, alcohol, etc.)
Fuel pressure
Exhaust leaks
Problems in the air pump system

If all of these items check out okay then replace the oxygen sensor.

Inspect the purple wire to the ECM for a short to ground, if the wire is okay replace the ECM.

control terminal (E9). If there is voltage at the EGR but not at the ECM, check for an open circuit in the EGR control windings. There should be less than 20 ohms. If there is more, replace the EGR valve. If the EGR is okay, repair the gray wire from EGR terminal A to ECM terminal E9. If there is not 12 volts at the pink and black wire at the EGR, check for a blown gauges fuse. If the fuse is okay,

repair the open circuit in the pink and black wire to the EGR valve (CKT 39).

If there is no vacuum supplied to the EGR valve, check and repair the vacuum supply hose. If the supply of ported vacuum is okay, replace the EGR valve.

If there is a pulse on the EGR control terminal, and if there is vacuum to the EGR valve but the voltage at the EGR position terminal is between 0.2 and 4.0 volts and not varying, replace the EGR valve. If the voltage is below 0.2 and no other codes are set, repair the open circuit in the red wire (CKT 357) connected to terminal A3. (If other codes are also set, make those diagnoses first.) If the voltage is greater than 4.0 volts, repair the short to voltage in the red wire, CKT 357.

Type D

Type D is the simplest of the Code 32 variations to troubleshoot. This type of system is used on the TBI-equipped cars such as the 2.0- and the 2.5-liter engines.

Replace the EGR valve if it feels loose. Check for proper ported vacuum supply to the EGR valve, and repair as necessary.

Connect a hand-held vacuum pump to the EGR valve, then pump up 15 inches of vacuum. If the EGR valve does not hold vacuum, replace the EGR valve.

Code 33

Code 33 refers to the sensor that is used to measure the mass or volume of the incoming air. Whether the device used on a given application is a Bosch MAF, a Delco MAF, or a MAP sensor, Code 33 always refers to the sensor signal being high. In the case of the Delco MAF it refers to a high frequency. In the case of the Bosch MAF and the MAP sensor it refers to a high voltage. Which one of these the code refers to will depend on which one the car is equipped with. If the car has both a Delco MAF and a MAP, then the Code 33 refers to the MAF.

Delco MAF

Clear the codes, start the engine and allow it to run until the Check Engine light comes on or until you are satisfied that it will not.

If the Check Engine light comes back on, connect a digital tachometer (set on the four-cylinder scale) to the yellow MAF signal wire at the MAF. Turn the ignition switch on without

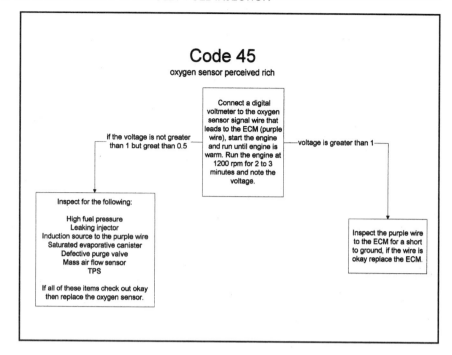

starting the engine. If the tach reads more than 4,000 rpm, replace the MAF. If it reads less, the problem is intermittent and no repair is needed at the time. (The most probable repair is still replacement of the MAF.)

Bosch MAF

Clear the codes and run the engine until the Check Engine light

comes back on or until you are satisfied that it will not.

Since this code indicates that a high-voltage signal has been seen on the MAF signal wire, the possibilities are an open in the MAF signal wire between the ECM and MAF, a bad connection at the ECM, a bad MAF, or a problem with the power supply to the MAF.

If the Check Engine light comes back on, connect a digital voltmeter to the MAF output wire, dark green CKT 998 at the MAF. Turn the ignition switch on. If the voltage is close to 0, then check the voltage at terminal B12 of the ECM. If the voltage at B12 is 5 volts, repair the open circuit in the dark green wire, CKT 998, which runs from the ECM to the MAF.

If there is 0 volts at terminal B12, inspect the connection to the ECM. If the connection is good, replace the ECM.

When checking voltage at the dark green signal wire of the MAF if the voltage is 5 volts, check the connection to the MAF. If the connection is good, check for a 12-volt power supply to the MAF on the purple wire with the engine running. If there is 12 volts, check the ground at terminals A and B of the MAF. If the ground is okay, replace the MAF.

If there was 0 volts on the purple wire at the MAF, apply 12 volts to terminal G of the ALCL connector. If 12 volts now appears on the purple wire, check the fuel pump fuse. If the fuse is okay, replace the oil-pressure sending unit. If the purple wire still does not have 12 volts on it when power is applied to terminal G of the ALCL, check the wiring to the MAF power relay. If the wiring is good, replace the MAF power relay.

MAP

Clear the trouble codes and run the engine until the Check Engine light comes back on or until you are satisfied that it will not.

Connect a digital voltmeter to the MAP signal wire (usually light green) to the ECM. Turn the ignition switch on; do not start the engine. If the voltmeter reads over 4.8 volts, inspect the MAP signal wire for a short to voltage. If the wire is okay, replace the MAP.

Code 34

Code 34 is the opposite indication from a Code 33. With this code the ECM has perceived a low frequency from a Delco MAF, a low

Code 52

MEMCAL or PROM fault detected. Inspect that all the pins are fully inserted. If they are replace the MEMCAL for PROM. If code 52 returns after replacing the chip replace the ECM.

Code 55

Replace the ECM.

voltage from a Bosch MAF, or a low voltage from a MAP. Keep in mind that for either one of the MAF sensors, Code 34 can be set due to a false air leak.

Delco MAF

Clear the codes and run the engine until the Check Engine light comes back on or until you are satis-fied that it will not.

If the Code 34 returns, connect a digital tachometer to the yellow MAF signal wire at the ECM. Does the tach (on the four-cylinder scale) indicate a frequency greater than 960 rpm? If it does, replace the ECM.

If the tachometer does not indi-cate a frequency greater than 960 rpm, disconnect the MAF connector.

In the key-on and engine-off mode, the yellow wire should have 5 volts, the pink and black should have 12 volts and the black and white should have a good ground. Repair any wire that does not have the proper voltage. If the voltages are correct, replace the MAF.

Bosch MAF

Clear the codes and run the engine until the Check Engine light comes back on or until you are satisfied that it will not.

If Code 34 returns, disconnect the MAF and run the engine for 2 or 3 minutes. Does a Code 33 set? If it does not, the MAF signal wire (CKT 998) is shorted to ground or the ECM is faulty. If the Code 33 is set, check for 12 volts available to the MAF sensor. If the MAF is receiving 12 volts, replace the MAF.

MAP

Clear the codes and run the engine until the Check Engine light comes back on or until you are satisfied that it will not.

If the Code 34 resets, and there are no other codes, connect a digital voltmeter to the MAP input terminal of the ECM. If the voltmeter displays between 0.5 and 4.0 volts, check the ECM connection. If the connection is good, replace the ECM.

With the key on and the engine off, disconnect the MAP harness and test the gray 5-volt reference wire. If the reading is not 5 volts, repair the open circuit between the harness and the ECM. If there is 5 volts, place a jumper wire between the 5-volt reference wire at the harness connector and the MAP signal wire. If the voltmeter displays 5 volts, check the MAP signal input connection at the ECM. If the connection is good, replace the MAP.

Code 35

Idle speed error. Allow the engine to idle until you are sure that operating temperature has been reached. Note the rpm, shut the engine off and wait 10 seconds.

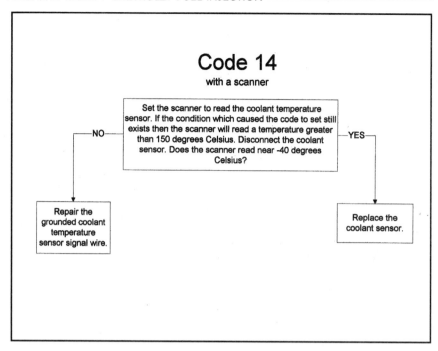

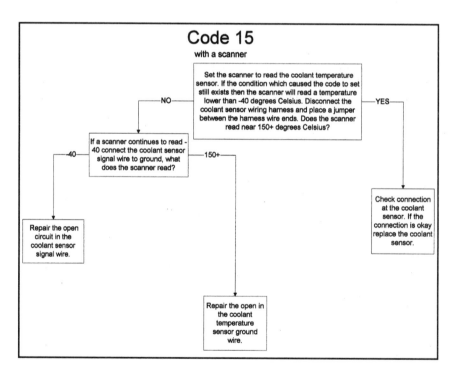

Restart the engine and immediately note the rpm. The rpm should be higher than that noted before the engine was shut off. Allow the engine to idle for 1 minute. The idle should return to close to what was noted before the engine was shut off. If all these conditions are met, the problem no longer exists.

If the idle does not perform as

described, connect a test light to all four of the IAC terminals one at a time and observe. If the light blinks on and off on each of the terminals, check the IAC connection. If the connection is okay, replace the IAC.

If any of the terminals do not blink on and off, check for an open, short, or ground in the circuit that did not blink.

If the circuit is okay, check the resistance between terminals A and B and terminals C and D. If the resistance is less than 20 ohms on either set of windings, replace the ECM. If it is greater than 20 ohms, check the IAC connections at the ECM. If the connections are okay, replace the ECM.

Code 36

When a 5.0- or 5.7-liter Tuned Port Injection car is shut off after being run in closed loop, the heated wire in the MAF is heated red-hot to clean off anything that may have stuck to it while the engine was being run. This code relates to a failure in the Mass Air Flow burn-off circuit.

Clear the codes. Start the engine and allow it to run for several minutes to ensure that it has gone into closed loop. Turn the engine off and wait about 20 seconds; then restart the engine and allow it to run about 30 seconds or until the Service Engine Soon light comes on. Does the Code 36 reset?

If it does not, the problem is intermittent and no repair is indicated at this time.

If Code 36 returns, probe both orange battery-feed circuits (MAF burn-off relay terminals D and E) with a test light. If either wire does not have power even with the key off, repair the open circuit in that wire to the battery.

If both wires light the test light, disconnect the MAF and ground the black wire going into terminal D12 of the ECM. Probe the D terminal of the burn-off relay with the test light. Is the light on?

If the test light does not light up, find and repair the open circuit in the black wire to terminal D12 of the ECM. Also inspect the dark blue wire into the MAF burn-off relay for an open, short, or ground.

If the test light does light up, keep the black wire grounded and probe the red wire in the connector that would normally be connected to the MAF. If the test light does not light up, repair the open circuit in the red wire. If the red wire does light the

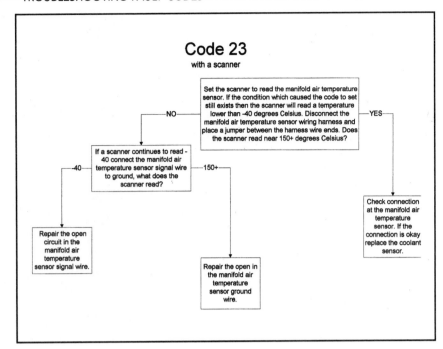

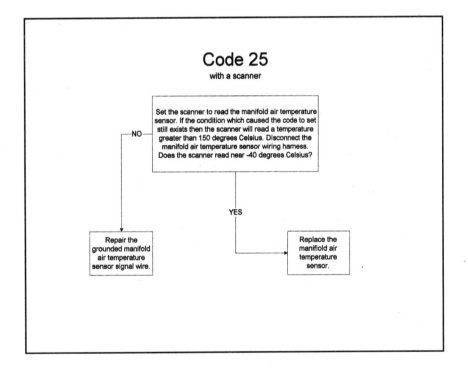

test light, then check for a short to power on the black wire. If the black wire is okay, replace the ECM.

Code 41

Cylinder select error. If the calibration chip has been recently replaced, this code has been set because the programming is incorrect for the car. If it has not been

replaced recently, then check for signs of corrosion. If there is no corrosion, replace the calibration chip.

Code 42

Clear the codes, then run the engine to confirm that the condition that caused the code to set still exists. If the code returns, probe the tan and black by-pass wire with a voltmeter at the ECM. If the voltmeter reads 0 volts, disconnect the timing set connector in the tan and black wire and note the voltage. If the voltmeter continues to read 0 volts, then inspect the tan and black wire back to the ECM. If the wire is okay, replace the ECM.

If, when you disconnect the set timing connector, the voltage increases to 5 volts (with engine still running), inspect the tan and black wire to the ignition module. If the wire is good, replace the ignition module.

When you originally probed the tan and black wire with the engine running, if there was 5 volts on the wire, connect a digital dwell meter to the white EST wire and note the dwell reading. If the dwell meter reads anything other than 0 or 90 on the four-cylinder scale, rev the engine. If the dwell changes, connect a timing light to the engine and note timing changes as the engine is revved. If the dwell changes and the timing changes, replace the ECM. If the dwell meter reads 0 or 90, replace the ignition module.

Code 43 (with ESC Module)

After confirming that the condition that caused the code to be generated still exists, check for 12 volts on the pink and black wire that is connected to the ESC module with the key on. Repair the wire or replace the fuse as necessary.

If the pink and black wire has 12 volts on it, turn the ignition switch off and connect an AC voltmeter to the white wire of terminal E on the ESC module. While closely watching the AC voltmeter, tap on the engine block with a 4-ounce hammer. Does the voltmeter indicate any reading? If it does, then the knock sensor is good. If it does not, replace the knock sensor.

If the knock sensor tests positively, connect a digital logic probe to the ESC terminal of the ECM. If it reads digital low with the engine running, inspect the black wire from the ESC module to the ESC terminal of the ECM. If the wire is okay, replace

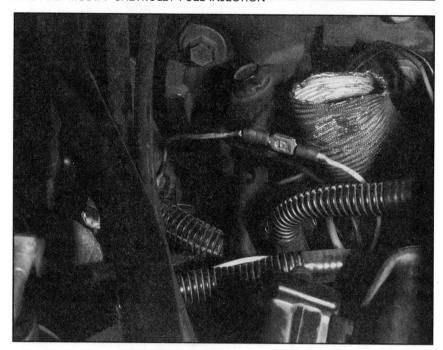

When repairing a fuel-injection system wire, always solder the connection. Crimped connectors do not always make an adequate connection. Having said that, the author deliberately cut this wire over two years before this photo was taken, and it has never caused a problem.

the ESC module. If there is a digital high on the ESC terminal of the ECM, tap on the engine block. If the logic probe fails to latch for a digital low, then replace the ESC module.

Integrated Knock Sensor

Disconnect the knock sensor wiring connector. Check the voltage on the wire going back to the ECM. If there is 5 volts on this wire, replace the knock sensor. If there is 0 volts, inspect the ECM knock sensor input wire for a short or ground. If the wire is okay, replace the ECM.

Code 44

Code 44 is set when the oxygen sensor indicates a lean condition. Connect a digital voltmeter to the purple oxygen sensor signal wire of the ECM. Start the warm engine and run at 1,200 rpm for 2 to 3 minutes while observing the voltmeter. If the voltmeter reads 0 volts (or very close to 0), look for a short to ground in the purple oxygen sensor wire to the ECM. If the purple wire is not grounded, replace the ECM.

If the voltage is not at 0 but remains consistently below 0.35 volts, inspect the following: MAF sensor and false air, restricted injectors, contaminated fuel (water, alcohol and so on), fuel pressure, vacuum leaks, exhaust leaks and problems in the air pump system.

If all of these items check out okay, replace the oxygen sensor.

Code 45

Connect a digital voltmeter to the ECM terminal for the purple oxygen sensor input wire. Run the warm engine at 1,200 rpm for 2 to 3 minutes and note the voltmeter readings. If the voltmeter is cycling throughout the entire oxygen sensor output range, the problem is intermittent and no repair is indicated at this time. If not, disconnect the oxygen sensor and ground the purple wire. If the voltmeter reads greater than 0.35 volts, inspect the purple wire for a short to voltage. If the purple wire is okay, check the following: fuel pressure, leaking injector, induction source to the purple wire, saturated evaporative canister or defective purge valve, MAF, and TPS.

If all of the above test okay, then replace the ECM.

Code 51

Code 51 says a MEMCAL or PROM fault was detected. Inspect that all of the pins are fully inserted. If they are, replace the MEMCAL or PROM. Start the engine and monitor the Check Engine light. If Code 51 returns after replacing the PROM or MEMCAL, replace the ECM.

Code 52

This error indicates a CALPAK error has been detected. Inspect that the CALPAK is properly installed in the socket. If properly installed, replace the CALPAK. Start the engine and wait for the check engine light to come back on. If Code 52 resets, replace the ECM.

Code 55

If a Code 55 occurs, replace the ECM.

Troubleshooting with the Scanner

Some of the fault codes are much easier to troubleshoot with the use of a diagnostic scan tool, especially codes 14, 15, 23, and 25.

Code 14

Set the scanner to read coolant temperature. If Code 14 is a hard code, the scanner will read greater than 302 degrees Fahrenheit. If it does not, then the problem that caused the code to set no longer exists. Disconnect the coolant sensor. If the scanner still reads 302+ degrees Fahrenheit, repair the ground coolant sensor signal wire.

If the scanner reads -40 degrees Fahrenheit when the sensor is disconnected, replace the sensor.

Code 15

If the problem that caused the code to be set still exists, the scanner will read -40 degrees Fahrenheit or so. Disconnect the coolant sensor harness and replace a jumper across the terminals.

If the scanner reads 302+

degrees Fahrenheit, inspect the coolant sensor connection. If the connection is okay, replace the coolant sensor.

If the scanner continues to read -40 degrees Fahrenheit, connect the coolant sensor signal wire to ground. If the scanner still reads the same, repair the coolant sensor signal wire. If the scanner reads 302+ degrees Fahrenheit, repair the coolant sensor ground wire.

Code 23

If the problem that caused the code to be set still exists, the scanner will read -40 degrees Fahrenheit or so. Disconnect the MAT sensor harness and replace a jumper across the terminals.

If the scanner reads 302+ degrees Fahrenheit, inspect the MAT sensor connection. If the connection is okay, replace the MAT sensor.

If the scanner continues to read -40 degrees Fahrenheit, connect the MAT sensor signal wire to ground. If the scanner still reads the same, repair the MAT sensor signal wire. If the scanner reads 302+ degrees

Fahrenheit, repair the MAT sensor ground wire.

Code 25

Set the scanner to read MAT temperature. If Code 25 is a hard code, the scanner will read greater than 302 degrees Fahrenheit. If it does not, then the problem that caused the code to be set no longer exists. Disconnect the MAT sensor. If the scanner still reads the same, repair the ground MAT sensor signal wire.

If the scanner reads -40 degrees Fahrenheit when the sensor is disconnected, replace the sensor.

On-Board Diagnostic System Generation 2 (OBD-2), the universal code system mandated by the Environmental Protection Agency, uses a different set of codes than older OBD systems. In chapter 14, there is a list of these codes. The only significant changes, from a troubleshooting perspective, are the diagnostic codes. The circuit designs and the methodologies do not change. Compare this list of codes and their definitions to the list of code definitions in this chapter, or use the chart below.

If the code is related to:	Then use this troubleshooting routine:
Sluggish oxygen sensor	Code 13
High coolant temperature indicated (low voltage)	Code 14
Low coolant temperature indicated (high voltage)	Code 15
High TPS voltage	Code 21
Low TPS voltage	Code 22
High air temperature indicated (low voltage)	Code 23
Vehicle speed sensor inoperative	Code 24
Low air temperature indicated (high voltage)	Code 25
Computer outputs	Code 26
EGR related	Code 32
MAF signal high	Code 33
MAF signal low	Code 34
Idle speed problems	Code 35
PCM programming error	Code 41
Timing control problems	Code 42
Knock sensor related problems	Code 43
Oxygen sensor lean	Code 44
Oxygen sensor rich	Code 45
PCM problems	Code 52
PCM problems	Code 55

C
O
N
T
E
N
T
S

STREET PERFORMANCE 140

CAMSHAFT 145

CYLINDER HEADS 146

VALVE ACTION 146

EXHAUST SYSTEM 147

COMBUSTION CHAMBER CHARGING 148

FUEL-INJECTION SYSTEM MODIFICATIONS 149

FUEL SUPPLY MODIFICATIONS 149

ELECTRONIC CONTROLS 151

MODIFYING OBD-2 ENGINES 153

This formula can be used to estimate the expected cubic feet per minute (cfm) of airflow into a given engine at a given rpm. For example, using this formula yields an estimated airflow of 124 cfm on a 2.5-liter engine at 3,500 rpm.

CHAPTER ELEVEN

FUEL INJECTION PERFORMANCE MODIFICATIONS

A Nearly 20 years have passed since the golden age of the Detroit performance car. For nearly two decades, concerns about emissions, fuel economy, and safety have condemned the enthusiast to either late-model slug-bait or rebuilding wrecking yard refugees. However, in the latter part of the 1980s and heading into the 1990s, the late-model sleeper is again a realistic possibility.

The manufacturers have shown a renewed interest in performance, starting with the 1984 Buick Grand National. In 1987 Oldsmobile introduced the Quad 4 engine. This was a four-valve-per-cylinder, 2.3-liter four-cylinder engine that produced 154 hp stock—more than 1 hp per cubic inch. Companies that were performance names in the 1960s, along with a whole basket of new companies, are hitting the market with performance parts.

There is one major difference between modifying engines from back in the 1960s and modifying engines today—emission laws! Different jurisdictions will have different mandates concerning emission control systems and what modifications are legal. As a whole, throughout the United States it can be safely said that it is illegal to remove or modify any emission control device on a car that will be licensed for street use. There is, however, a small amount of latitude permitted in some states for the replacement of some components like intake manifolds, throttle bodies and camshafts with specific tested and approved parts.

This chapter will therefore cover two major categories: Street performance modifications, and racing modifications. Please keep in mind that regulations and approvals change frequently when it comes to emission controls, so be sure to ask your parts supplier if a given modification is street-legal in your jurisdiction. Do not accept what this book describes as a legal modification to necessarily be legal where you live.

Street Performance

Any real increase in performance has to begin with an improvement in the engine's ability to breathe. We are going to divide the improvement of engine breathing into six categories: intake; camshaft; cylinder head; valve action; exhaust system; and combustion charging.

Intake System

For street use, many of the modifications to the intake system that were traditional in the 1960s have already been done by the manufacturer when the engine was designed. The simple fact is that any meaningful modification is going to be found either in modifications to the air cleaner and ducting to the throttle body, or in the complete rethinking

```
FORMULA FOR AIR FLOW

Cubic inches of the engine
───────────────────────────   =Cubic feet
          1728

Cubic feet  X  RPM
──────────────────           =Ideal CFM
        2

Ideal CFM X .8               =Actual CFM
```

Although the stock air cleaner fitted to a production engine has an acceptable flow rate for normal usage, anyone considering major engine performance modifications should replace the air cleaner with a high-performance model.

and reworking of the intake side of the engine including throttle body, intake manifold, and valvetrain.

Let's begin by looking at what you already have on the car and what can be done to improve it. The example we are going to be looking at is the 5.7-liter Tuned Port Injection Corvette V-8 engine. As we are looking at each of the topics, we will look at the availability of over-the-counter performance parts for not only the 5.7- but also the 5.0- and 2.5-liter engines.

Let's begin with a little mathematics relating to the air requirements of the engine. The air demands of the stock 5.7-liter (350 ci) Chevrolet small-block V-8 engine can be estimated with the following formula:

$$\underline{\text{Engine ci}} = \frac{350}{1{,}728} = 0.2025$$

This formula converts the standard measurement for engine size (cubic inches) into the standard measurement for airflow (cubic feet). The result of 0.2025 describes the volumetric change of all the cylinders as the pistons move through one complete engine cycle. Since all of the engines that Chevrolet uses are four-stroke engines, one engine cycle is represented by two revolutions of the crankshaft.

From this figure of cubic feet of displacement, we can derive the cubic feet of air inhaled by the engine per minute (cfm) by multiplying the cf by the maximum engine rpm we plan on running. We will then divide this number by two, since each cylinder inhales air only on every other crankshaft rotation. For the purpose of discussing airflow we will limit the rpm to 5,250.

$$\frac{\text{cf x rpm}}{2} = \frac{0.2025 \times 5{,}250}{2} = 532$$

At 5,250 rpm the maximum amount of air that can be pumped through the 350-ci engine is 532 cfm. This figure is not realistic, however, because it assumes 100 percent volumetric efficiency. In other words, it assumes that each cylinder at this speed is capable of being *completely* filled with air up to atmospheric pressure each time the intake valve opens. As we work to improve the performance of the engine, we will be working to increase the volumetric efficiency ever closer to 100 percent. Keep in mind, though, that in spite of product claims of power increases and bolt-on horsepower, it is a violation of some of the basic laws of physics to expect more than 100 percent volumetric efficiency without pressurizing the combustion chamber with a supercharger or turbocharger. This limits our ability to increase performance simply through improving the intake system.

Removing the MAF screens and cooling fins on Tuned Port Injection applications can significantly increase airflow. Flow bench testing has shown an increase of up to 182 cfm.

As would seem logical, the larger the throttle bore, the larger the potential airflow. Bolting on a new, larger throttle assembly may be the easiest way to get a significant increase in performance.

For a stock 350-ci engine, it would be fair to assume 80 percent Volumetric Efficiency (VE). This means that the actual cfm inhaled by our example engine will be lower than 532 cfm.

Ideal cfm x VE = Actual cfm (estimated)
532 x 80 percent = 426 cfm

Since the expected airflow through the stock engine is 426 cfm, we must be able to flow at least this much through all of the intake components just to avoid limiting power.

Let's look at how the Corvette air induction system can be improved at a relatively low expense. The source for a lot of the following is flow bench research done by TPI Specialties of Chaska, Minnesota (see appendices for address).

Air Cleaner

During normal operation, the air cleaner on most cars is capable of flowing more air than the engine actually requires. If the car is to be used under circumstances where maximum performance is required, such as racing or trailer towing, then the air cleaner assembly can become a restriction to airflow and therefore performance.

In the old days, guys would flip the air cleaner lid upside down. While the effects of doing this were largely psychological, it did bypass the snorkel tube of the air cleaner and greatly increase the potential airflow. The design of fuel-injection air cleaners does not make inverting the lid practical or desirable.

According to TPI Specialties' airflow studies, the air cleaner with a stock lid and a stock air filter is capable of flowing 648 cfm. This seems to be a comfortable amount above the 426 estimated cfm for the 350-ci Tuned Port Injection engine, and indeed modification on a stock engine would not be necessary. In fact, there is enough airflow capability in the stock air cleaner to handle 6,400 rpm at 100 percent volumetric efficiency. What this means to the average motorist is that the air cleaner assembly will become a factor only when there is a desire to suddenly change the pressure in the intake manifold.

When the throttle plates are opened, the first thing that occurs is a sudden increase in the mass velocity of the air entering the intake manifold. Cruising at 2,000 rpm, our sample 350-ci engine is swallowing only about 141 cfm. When the throttle is flopped wide open, the demand suddenly jumps to a potential of 434 cfm, and as the rpm climbs so will the demand. The more open and free flowing the intake system is ahead of the throttle plates, the faster the air mass will build and therefore the faster the torque will build.

TPI Specialties and others make air cleaner housings for the 5.7- and 5.0-liter TPI applications that boast increased cfm capability and therefore an increased ability to make a sudden change in mass airflow rate. A crude but effective method of accomplishing the same thing is to trim the air filter lid so that all of the air filter is exposed, thereby eliminating the air cleaner's lid as a restriction to airflow through the filter.

Air Filter

The air filter itself can severely impede airflow. K&N, TPI Specialties, Hypertech, and other companies offer high-flow air filters for the 5.0-

and 5.7-liter engines. The filter costs about $50 but is one of the easiest and most legal ways of improving power, and since they can be cleaned and re-used, they can last for years. Hypertech also offers performance air filters for V-6 and four-cylinder applications, while K&N offers performance filters for almost any engine.

How do these performance air filters differ from a standard air filter to the extent that there can be a noticeable improvement in power? To understand that, we need to begin with the premise that a stock air filter is designed to be easy to use and inexpensive to produce. The consumer that is not performance oriented is interested in reducing the cost of maintenance, and these stock-style air filters are designed to meet that need. A performance air filter sacrifices low price for airflow.

The high-performance air filter consists of multiple layers of cotton gauze that has been treated with a light oil. The oil attracts the dirt and dust particles, allowing the gauze element to have a more open weave yet still have good filtering capability. Typically the filtering element will be layered across a metal grid which holds the gauze in an accordion shape. As this accordion shape weaves its way back and forth, the surface area for the passing of air into the engine is greatly increased. High-performance filters can result in a 25 percent increase in cfm through the filter. Keep in mind that a filter capable of flowing 500 cfm will not increase performance significantly if it is mounted in an air cleaner assembly that will only flow 200 cfm.

Some of the sales literature concerning performance air filters states that they are not legal for use on California pollution-controlled cars. The 1990 BAR (Bureau of Auto Repair) *Smog Check* manual states in Appendix K that the air filter is a Category 1 component. Category 1 replacement parts are not considered to be "of concern" as long as none of the emission control devices themselves are tampered with during the installa-

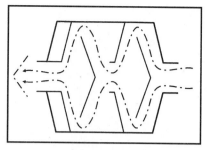

The modern high-tech performance muffler, such as those from Flowmaster, is free flowing, using the programmed collision of frequencies to cancel sound. The back-pressure that does exist in the muffler aids in the scavenging of the cylinders and actually increases power.

tion. Check with the shop that does your smog check about their interpretation of this rule, or contact your local BAR field office for a ruling.

It's important to understand that while these $50 filters are designed to be a lifetime replacement, they do need to be serviced on a regular basis. A special solvent or cleaner is sold that removes road oil and dirt. You will find recommendations on the correct cleaners to use in the literature that comes with your new air filter. After cleaning, the oil barrier must be replaced. Using the wrong oil—penetrating oil for instance—can reduce the efficiency of the air filter to the point where you would be better off with a stock filter. Again, follow the recommendations of the filter manufacturer. Plan on cleaning the air filter every time you change the engine oil. This may be a little more frequent than is really required, but it can't hurt.

Mass Air Flow Sensor

One of the worst offenders in the restriction of airflow is one of the most necessary devices in the entire fuel-injection system—the Mass Air Flow sensor. The MAF used on our TPI Corvette was designed and built by Bosch, and according to TPI Specialties' studies is capable of flowing only 529 cfm stock. The MAF's greatest restriction are the screens located at each end of the sensor that protect the delicate heated wire suspended across the main channel of MAF. Flow bench testing indicates that an increase of 182 cfm—upping flow to a respectable 711 cfm—can be obtained by simply removing these screens.

The MAF also has cooling fins for the electronic module on the side of the sensor that sit across the main channel of airflow. These fins represent a restriction to airflow of about 39 cfm. While this is not nearly the restriction created by the screens, these cooling fins represent a potential loss of about 10 percent of the total horsepower potential of the engine. Not only do we have these fins to contend with, but we also have the fact that the hot wire sensing element sits across the main channel of airflow, seriously reducing the potential airflow as a function simply of the unit doing its job. A rough estimate of the restriction created by the sensing element is about 20 percent, and while we cannot do anything about the sensing element, we can do something about the cooling fins.

Legend has it that electronic failures of the MAF occurred during Death Valley testing. These cooling fins were then added to increase the ability of the MAF electronics to dissipate heat in high-temperature environments. There is probably a certain amount of truth to the legend, since the Bosch fuel-injected applications such as Volvo and Porsche sport no such large fins.

If you are not going to be operating the car in a hot-weather environment, then grab your hacksaw, your 200 mile an hour tape and get busy. Cover each end of the sensing element venturi (the circle within the circle) with the tape; ensure that the interior of the sensor is free of grease, oil, moisture, or anything else that might attract and hold the metal sawdust that is going to result from the removal of the cooling fins.

If you are like me, you may have something ranging from quiet trepidation to out and out fear about taking a hacksaw to a $400-plus electronic component. For people like

Turbochargers and superchargers increase horsepower by compressing the air in the cylinder. There are several excellent books on the market that cover turbocharging and supercharging in detail.

me, TPI Specialties sells a stock unit to replace the one I screw up, or one that is already modified for less than dealer list. If you prefer, and can do without your car for a few days, your MAF can be professionally modified for less than $150.

The PFI application uses the Delco MAF. The design of this unit does not lend itself to modification.

Oddly, Oldsmobile's 3800-series engines use a Japanese-designed MAF. In this Hitachi design, the hot-wire sensing element is located in a bypass tube outside the main channel of airflow. This technology not only eliminates the airflow restriction created by the cooling fins, but also the restriction of the sensing element itself, leaving, in effect, a big hollow tube for the air to pass through. Unfortunately for the performance enthusiast, this MAF is not interchangeable with the Corvette MAF. (That is, unless you possess an exceptional amount of free time and a degree in electronic engineering.)

Throttle Body Assembly

The throttle body assembly (not to be confused with Throttle Body Injection units) holds possibly the most important potential for bolt-on improvement, short of the intake manifold itself. It is important to stress again for this application that Chevrolet spent much time and effort creating the TPI system to be a good performer right off the showroom floor. As a result, any improvement in performance afforded by any of the techniques discussed here, including replacement of the throttle body may be less likely to show up in the seat of the pants than it is on the quarter-mile. Even quarter-mile results might prove disappointing to all but the most die-hard racer.

Several companies manufacture high-performance throttle body assemblies, but let's begin our throttle body modifications by looking at an inexpensive and clever approach. TPI Specialties and Hypertech sell an airflow directing device that helps to guide the incoming air through the throttle bore and into the intake manifold. Called an airfoil, this unit simply bolts on to the throttle assembly and adds an additional 8–10 hp. Documentation on the 350-ci Corvette engine shows an increase of up to 13

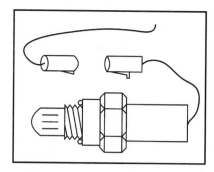

Any modifications to the fuel supply system will be counteracted by the oxygen sensor. Therefore, disconnect the oxygen sensor whenever fuel pressures are altered.

additional lb-ft of torque at 4,500 rpm. Fine tuning the flow of air through the throttle-valve assembly with a device such as the airfoil increases cfm in a flow bench test by 17.

Big-bore kits are available from Air Sensors and others. These kits consist of a complete replacement throttle-body assembly. The Air Sensors version features a single, large throttle plate replacing the small pair of throttle plates in the stock unit.

Holley markets a wide range of TBI replacement units. These replacement units fall into two general categories: Stock Emission Legal/Street Performance; and Street Competition/Non-emissions-legal.

The emission-legal units are available in a 300 cfm throttle-body assembly for the 2.0- and 2.5-liter applications, and a 400 cfm two-barrel version for the 2.8-liter V-6 TBI truck engines. These throttle bodies are direct replacement bolt-on assemblies that come complete with the TPS and IAC. Once installed, all you need to do is connect the vehicle wiring harness to the TPS, IAC, and injector, adjust minimum air and TPS closed-throttle voltage, then drive the car away. They are for use on model years 1982 through 1986. Applications later than 1986 use a different throttle body assembly that is not interchangeable with these units.

The non-emission TBI units come with their own computer and wiring harnesses. These will be covered in chapter 13 on aftermarket fuel injection.

When the installation of either the airfoil, the big-bore kit, or replacement TBI is complete, it may be necessary to adjust the minimum idle speed. Unless specified in the installation instructions or in data supplied with other components that have been added to the engine, you should adjust the minimum idle speed to the specifications found in the Minimum Air and TPS Specifications chart (see appendices).

Intake Manifolds

The stock TPI manifold is designed for peak performance; however, production tolerances are not such that perfection is always achieved. The TPI manifold consists of a large central plenum box connected to the intake manifold base by individual tubes, called runners. Some improvement in cfm can be achieved by matching the sizes and alignment of the plenum, runners, and manifold base.

Matching the alignment of the plenums can be achieved by applying a light film of Prussian Blue to each end of the runner, and installing and torquing the runner down. Removing the runner and inspecting the Prussian Blue contact points will show where the runners are misaligned. A more careful inspection will reveal where the runners are undersized or oversized compared to either the intake or central plenum. A die-grinder and an ample supply of patience will enlarge the undersized orifices to match the larger. Do not get carried away, though. The size and shape of the plenum, runners, and manifold are all basically correct.

After painstaking hours of work you will have gained only about 2 or 3 hp. You might say that it is hard to improve on what is basically a good design.

There are high-performance manifolds available for the 5.0- and 5.7-liter engines from Edelbrock, TPI Specialties, Air Sensors and others. These manifolds offer a rethinking of the basic Chevrolet design in addition to larger porting, improved run-

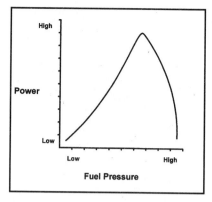

Increasing fuel pressure can have a dramatic effect on power up to a point. There are many variables that determine how high the pressure can be increased before the power curve starts downward—things like camshaft design, exhaust design, and other performance modifications.

ner matching, and improved runner flow angles. However, overall improvement in performance may be disappointing unless these manifolds are installed as part of a complete performance upgrade.

For the TBI applications, performance improvements from intake manifold modifications may be more noticeable. The *General Motors Performance Parts Catalog* offers performance intake manifolds only for use with a two-barrel or four-barrel carburetor. Holley sells a performance manifold for the 2.5-liter TBI engines.

Intake System Summary

Back in the 1960s, massive improvements in performance could be effected through relatively minor and often inexpensive modifications. In those days, the large brutish engines made by Chevrolet had tremendous unlocked potential. The 396 ci had a flow rate potential of 458 cfm, although factory-equipped with about a 650 cfm carburetor (depending on the year and exact application this varied). Installation of a carburetor with a greater flow rate would instantly unlock horsepower and torque hidden by the inability of the stock carburetor to fill the manifold with air quickly enough when the throttle was matted.

$$\text{Lb./Hour} = \frac{\textit{Maximum HP X Brake Specific Fuel Consumption}}{\textit{Number of Injectors}}$$

When major engine modifications are made, the stock fuel injectors may not have an adequate pound-per-hour flow rate. This formula can be used to determine the correct pound-per-hour flow rate for the horsepower you will be producing.

Today's cars have evolved from the 1960s through a period in the 1970s when anything and everything was done to maintain an acceptable level of performance while maintaining a legal emission level. During the late 1970s and early 1980s, the manufacturers began a quiet mutual challenge back toward performance. This time they did not have the luxury of simply bolting on a larger carburetor and dumping gallons of gas through the engine. As a result, today's intake systems are highly refined when they roll off the assembly line. Consequently, a lot of work on the manifold without considering other engine, computer, and sensor modifications may yield disappointing results. Again, it's hard to improve on a good design.

Camshaft

The basic function of the camshaft is to open the intake valve as quickly, as smoothly, and as far as possible. The cam then leaves it open as long as possible to allow atmospheric pressure to push air and fuel into the combustion chamber. The camshaft must then begin (along with the assistance of the valve spring) to allow the intake valve to close in enough time for the combustion chamber to be sealed as the piston begins to move up on the compression stroke.

Actually, the intake valve remains open for a short time as the piston begins to move upward on the compression stroke. Leaving the intake valve open briefly allows the velocity of the air traveling through the intake system to continue to cram air into the cylinder even as the piston begins to move upward. This effect is even more pronounced on engines equipped with high-rise intake manifolds such as on the TPI

applications. This is why increasing the airflow potential beyond the capability of the engine's volumetric displacement will result in improved performance. Less restriction in the intake system will increase the cylinder charging capability.

The camshaft must also open the exhaust valve as far, as fast, and as smoothly as possible. The opening of the exhaust valve begins as the piston nears BDC (Bottom Dead Center) on the crankshaft power stroke. This ensures that the exhaust valve will be completely open when the piston begins to move upward on the exhaust stroke. The camshaft begins to close the exhaust valve near TDC (Top Dead Center); however, the exhaust valve remains open for a short time as the piston begins its downward travel.

This movement does two things: First, it takes advantage of the velocity of the exhaust gases exiting the combustion chamber to accelerate the incoming air-fuel charge. This increases the volumetric efficiency of the engine. Second, the other reason for leaving the exhaust valve open as the intake valve opens is that as the incoming charge is accelerated some of the outgoing exhaust gases will be slowed, causing the exhaust gases to be left in the combustion chamber. From the perspective of performance this is neither desirable nor efficient; however, it does produce an EGR effect, causing combustion temperature to be lower, thereby reducing the output of oxides of nitrogen (NOx).

Like the intake manifold, the TPI camshaft was designed by Chevrolet with performance in mind. Multipoint injection minimizes the emission considerations in camshaft design, limiting consideration to performance and driving comfort. As a

result, a significant performance improvement will come at a cost of low-speed, low-load drivability.

Several companies offer off-the-shelf camshafts for performance and economy. For obvious reasons, these performance cams are pretty much limited to the bigger engines—the 5.7-, 5.0-, 4.3-liters and the 60-degree V-6s. Designing a performance cam that will effectively increase torque and horsepower for an electronic fuel-injected engine is not a job for mere mortals. Changes in lift, duration, centerline, and timing that have been infallible in the past can cause a loss of power, under-fueling and overfueling today. The reason is, changing the camshaft will change intake manifold vacuum (pressure) and airflow rates. These changes will affect the readings of the MAP and MAF sensors. Changes in the MAP or MAF sensor readings at a given rpm or engine load may cause undesired changes in air-fuel ratio and ignition timing.

Crane Cams (see appendices for address) and others have lines of camshafts that have been designed specifically for use with electronic fuel injection. Because of the research and development that has gone into these cams, it is unlikely that a custom-ground camshaft, unless it is ground by the best of the best, would work as well.

While a local Joe, Super Racer custom-ground camshaft will provide for greater increase in performance, their added expense is not really justified for the typical performance enthusiast. Several grinds are available from different camshaft manufacturers. A typical and effective grind for the 5.7-liter Corvette engine might be as follows:

Exhaust
duration: 262 (hydraulic flat tappet)
Intake
duration: 266 (hydraulic flat tappet)
Lift: 0.451 inches
Intake
centerline: 104

This grind can provide a 6 hp increase at 2,000 rpm and up to 43 hp more at 4,500 rpm while maintaining acceptable idle quality.

If you decide to perform a camshaft transplant, be sure to check with your supplier about how it will affect the street legality of your car. For instance, you should consider that the increased valve overlap of a performance camshaft will tend to increase hydrocarbon emissions at an idle. Since many states test emissions only at an idle, installation of such a cam may lead to emission test failure. The California BAR considers camshaft replacements (other than with a stock grind) to be an unacceptable modification.

Cylinder Heads

The stock 350-ci cylinder head may well be the single biggest candidate for improvement. The basic design of the 5.7-liter head dates back to the 1960s, and although there have been improvements and upgrades along the way, it still remains a rather old piece of technology when compared to the cylinder head of a Quad-4 or late-model Japanese engine.

To improve the cylinder head of the 5.7-liter TPI engine, you have two good ways to go. Many aftermarket companies make performance cylinder heads ready to be bolted on and driven away. These heads sell for $500 to $1,000 or more but can provide the most noticeable change in performance of anything yet discussed. In most jurisdictions, these cylinder heads meet emission requirements with ease.

The second option is to modify the existing cylinder heads. You may prefer to have a machinist do this, for it requires a minimum of manual skill but a maximum of knowledge on the subject. Let's take a look at some of the basics.

Most people who get involved in a "port-and-polish" job do not really have a good understanding of flow dynamics. The result is that many performance modifiers end up

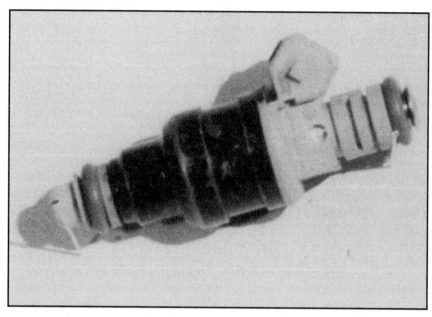

Although the PFI and TPI injectors all basically look the same, they will have different flow rates. If you are making performance modifications that involve significant horsepower increases, you should consider replacing the injectors with a set having a greater flow rate.

paying a lot of money or spending a lot of time to accomplish little or even to decrease performance.

The typical port-and-polish job consists of enlarging the intake and exhaust ports of the cylinder head and then polishing the enlarged surface to a high gloss. While this may look impressive, the stock ports are basically large enough for the camshaft applications found in common street use. Often the high-gloss shine reduces turbulence that can actually decrease the ability of the fuel to remain atomized in the air-fuel charge and therefore reduces performance. By the same token, for most street applications, large valves can have an adverse effect on airflow in and out of the combustion chamber.

The area that needs the most work or concentration is the valve pocket. This is the area just in front of the valve in the cylinder head. The valve guide boss can restrict airflow; on the stock 350-ci head, this boss is much larger than it needs to be. Additionally, there are casting ridges that can be removed by flattening and enlarging the radius where the air passage turns down toward the valve.

Shop around a little for a quali-

fied performance machinist if you decide not to do the work yourself, and remember that the best machinist may not always have the biggest reputation in town. If performance is really your goal, spend a couple of days at the local drag strip or circle track identifying the most consistently fast cars. Some of the fastest cars may not be consistent performers. Inconsistency may be a clue that the performance comes from bells and whistles rather than quality workmanship. Once you have identified the consistent performers, find out who does their machine work. This is the person you want working for you.

Just pick up a copy of one of the popular hotrodding magazines and you will be bombarded with ads offering custom-built small-block Chevrolet cylinder heads. These heads are just as valid for use on a TPI engine as they are on a carbureted engine. Most of these heads are well engineered and professionally done, making them well worth the money.

Valve Action

At this point I want to remind the reader that in this section we are talking about street performance, and

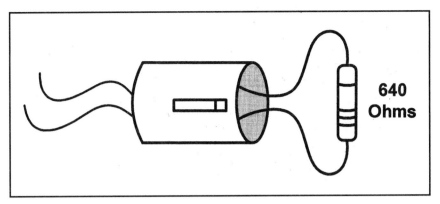

Placing a 640-ohm resistor across the terminals of the coolant temperature sensor harness tricks the ECM into believing that the engine is not quite warmed up yet. As a result, the ECM increases fuel flow into the engine and allows additional timing advance. The danger of this cheap trick is that some ECMs depend on the accuracy of the coolant sensor signal to accurately control the radiator fan. On these applications this trick could cause severe engine overheating and damage!

while there can be some tweaking such as cylinder heads, camshafts, and intake modifications, there are other performance parts that do not justify their expense on street application with improved torque or horsepower. One of these areas is the valvetrain. Where valvetrain modifications can provide an all-out race car with 0.1 second improvement on the quarter mile, these improvements would be seen only between the ears on a street rod.

Valvetrain modifications include higher-ratio and possibly roller rocker arms, high-tech pushrods, performance lifters, and heavy-duty valve springs. Engine rebuilders disagree on the necessity of replacing these components when the camshaft is replaced. When a new piece of metal rubs with a force of up to 300,000 psi against another piece of metal with only a thin film of oil in between (and we are going to expect it to do this for 100,000 to 150,000 miles) I would prefer that the new piece it is rubbing against also be new.

The bottom line is that if you decide to replace the camshaft, then spending a few extra bucks on high-performance valvetrain parts that should be replaced anyway makes good sense. Otherwise, put your money into areas where you can get more bang for the buck.

Exhaust System

Looking back, again, to the 1960s we saw a fascination among street enthusiasts with bigger pipes and freer exhaust flow. Today's stock exhaust systems are a far cry from the stock systems of the 1960s, though there is still room for improvement.

This improvement can begin with a set of headers bought over the counter with threads and fittings for the oxygen sensor and the air pump devices. Catalytic converters with high flow rates are also available. Be sure to check with local automotive emission officials concerning the legality of any exhaust modifications you are planning.

After the catalytic converter there is little that is not a legal modification in any state. Mufflers, connector and exhaust pipes are all fair game for the performance enthusiast. The only restrictions are related to noise.

Mufflers

Exhaust systems and their technology have come a long way since the muscle-car days of the 1960s. Today's high-performance mufflers exceed the flow potential of even open headers.

There are two things that travel down the piping from the exhaust manifold: Exhaust gases and sonic vibrations (or frequencies). The

movement of the frequencies through the exhaust system tends to pull the exhaust gases along in much the same way that ocean waves pull a surfer to shore. The effect of the exhaust gases being pulled through the exhaust system helps to scavenge the cylinder, improving the breathing of the engine.

Back in the sixties several companies marketed Turbo mufflers. This muffler was developed by Chevrolet to be used on the Corvair Turbo applications. It consisted of a hollow tube with fiberglass pressed against the sides of the tube to deaden sound. The problem with the Turbo muffler is that as the sound frequencies enter the muffler, they are killed by the fiberglass packing, negating the surfer effect.

Today's high-tech performance mufflers are able to reduce sound without eliminating the surfer effect. Imagine for a moment that you are setting up a stereo in your living room. The only place you can find to put one of the speakers is in the center of the north wall. The only place that you can find to put the other speaker is directly opposite the first on the south wall. The two speakers are facing each other squarely. The only place you can find to put your chair is exactly halfway between the two speakers. When the speakers are producing exactly the same sound, the frequencies being emitted from the speakers will collide and cancel one another, creating a dead zone.

Modern high-performance mufflers take advantage of this phenomenon. As the frequencies and exhaust gases enter the muffler, they are divided and sent in two different directions only to be brought back together as they pass through the muffler. When they are brought back together the identical frequencies that were split earlier collide and cancel each other like the speakers described earlier. The end result is less sound without a loss of the surfer action.

Currently, mufflers such as these are being used on everything from road racers to sprint cars.

H-Pipe Installation

Most exhaust systems have an area where these frequencies we have been discussing tend to build up and eddy the exhaust gases. To reduce this build-up on dual exhaust systems, it helps to install an H-pipe between the two sides of the exhaust. However, installing the pipe in the wrong place does more harm than not installing one at all.

To determine where an H-pipe is needed on your custom dual-pipe exhaust system, paint the area between the catalytic converters and the mufflers with black lacquer. Run the engine at 3,200 rpm for several minutes. Now inspect the painted area. Where the lacquer has begun to burn, or has burned the worst, indicates the place where the H-pipe needs to be installed. Install the pipe between the indicated hotspots in the two sides of the exhaust.

Combustion Chamber Charging
Turbochargers

One of the most impressive ways to increase the power from an engine is with the use of a turbocharger. For some applications there are even California-legal kits available.

A turbocharger uses hot exhaust gases to drive a turbine to extremely high speeds—on the order of 60,000 to 100,000 rpm. Sharing a common shaft with the turbine is a compressor that packs air into the intake manifold. Packing the air into the intake manifold forces the volumetric efficiency of the engine to above 100 percent.

The packing of the air into the intake manifold is called boost. Normally, boost is measured in psi gauge pressure. Gauge pressure is pressure above atmospheric pressure; therefore, when the boost is 15 psi, then the actual pressure in the intake manifold is 30 psi (15 psi boost plus 15 psi of atmospheric pressure). Significant horsepower increases can be achieved with even relatively small amounts of boost.

As an example, let's take a 1986

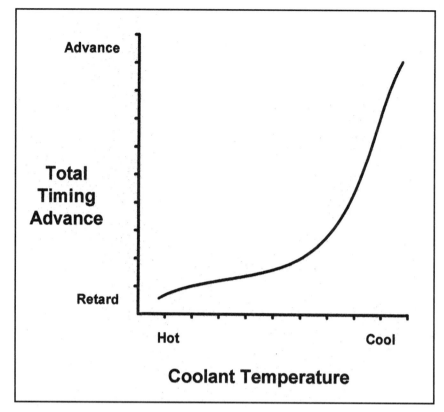

The power thermostat is a cold thermostat that legitimately keeps the engine cooler, allowing a greater intake manifold air density. Additionally, the ECM enriches the mixture and advances the timing.

5.0-liter Camaro and install a turbocharger set to deliver 10 psi of boost. What follows is a mathematical estimate of the expected power increase. Cubic inch displacement is converted to cubic feet (since cubic feet is the industry standard for measuring airflow into the engine).

$$\frac{305 \text{ ci}}{1{,}728} = 0.1765 \text{ cf}$$

Since airflow and therefore potential horsepower varies from one engine speed to another, any horsepower estimates have to be based on a given engine speed. For the sake of this exercise we will select 4000 rpm:

$$\frac{0.1765 \text{ cf} \times 4{,}000 \text{ rpm}}{2^*} = 353 \text{ cfm}$$

On a four-stroke engine each cylinder draws in air only on every other crankshaft rotation, so we divide by 2.

This figure of 353 cu-ft assumes a 100 percent volumetric efficiency, which as we have already stated is not possible on a normally-aspi-

rated engine. Again, a good approximation of volumetric efficiency is 80 percent.

$$353 \text{ cfm} \times 0.80 = 282 \text{ cfm}$$

Therefore, we can expect the cubic feet of air moving into the engine at 4,000 rpm to be about 282. For discussion we will say that this yields 200 hp on our engine.

When we boost the manifold pressure to 10 psi, the pressure inside the manifold will increase to a percentage above atmospheric pressure. This percentage is known as the pressure ratio. In the formula below, 29.92 is used to represent typical atmospheric pressure in inches of mercury (in. Hg.). The 2.03 is used to convert the standard boost readings of psi into inches of mercury.

$$\frac{10 (2.03) + 29.92}{29.92} =$$

$$\frac{20.30 + 29.92}{29.92} = \frac{50.22}{29.92} = 1.68$$

The 1.68 arrived at by the preceding formula is the pressure ratio.

Power chips replace the stock CALPAK. These chips significantly increase power by altering the timing advance curve and fuel enrichment specs. Be sure to check their legality in your area.

This means that the pressure in the intake manifold of our sample car is 68 percent above atmospheric pressure.

When a gas, such as air, is pressurized as we are describing in this example, its temperature increases. This process is known as adiabatic heating. The amount of heating is directly proportional to how much the pressure has been increased. A trip to your public library would yield a set of charts known as "Values of Y for Normal Air and Perfect Diatomic Gases." Using this chart you would find that for a pressure ratio of 1.68, the temperature increase would be by a Y factor of 0.15814. In other words, the temperature of the air compressed 68 percent higher than atmospheric normal will increase by 15.814 percent of the air's original temperature above absolute zero.

If the original temperature was 60 degrees Fahrenheit, then the original temperature was 520 degrees R (above absolute zero). That means that the temperature of the air has increased by 520 x 0.15814, or about 82 degrees simply by the act of being compressed. Assuming, again, that the original temperature of the air was 60 degrees, the temperature after compression is 142 degrees Fahrenheit.

Unfortunately, the temperature increase calculation is not quite that simple. Remember that what is driving the turbine is hot exhaust gases. Some of this heat will be transferred to the compressor side. This transfer of heat is known as compressor efficiency. Typical compressor efficiency is about 65 percent.

$$\frac{82° F}{0.65} = 126 \text{ degrees F}$$

This shows a 126 degrees Fahrenheit increase in temperature from both compression and heat transfer through the turbo. This makes the actual temperature in the intake manifold air temperature 186 degrees Fahrenheit. Right away you can see the potential for detonation due to the high intake temperature. But there is another negative effect from this increase in temperature.

As air is heated it expands, and as the air expands it becomes less dense. The less dense the air is, the less power there is in a cubic unit of air. The amount that the air has expanded in our example is shown with the following formula:

$$\frac{510 \text{ (ambient temperature in degrees R)}}{646 \text{ (manifold air temperature, degrees R)}} \times 1.68 = 1.35$$

The density of the air in the intake manifold at 4,000 rpm has increased from a normally aspirated 282 cfm to a boosted (282 x 1.35) 380 cfm. The horsepower increase will approximate the increase in true cfm, offering a potential increase from 200 up to 270 hp.

Superchargers

The act of supercharging an engine is far more complicated than turbocharging. There are drive belts and complicated piping to consider on fuel-injected cars.

A supercharger is like a turbocharger in that it increases power by pressurizing the intake manifold. It differs in that it is not powered by exhaust gases but by the engine itself. This has a tendency to rob some of the extra power it is creating just to drive the compressor. On the other hand, it does not have the problem with heat transfer that the turbocharger has, nor does it suffer "turbo lag" as the turbo's turbine begins to spin.

There are several kits available, including California-legal superchargers.

Fuel-Injection System Modifications

Important modifications to the fuel-injection system fall into two major categories: fuel supply system and electronics.

Fuel Supply Modifications

Some of the easiest and least-expensive performance modifications are those that can be made to the fuel supply system. Although the fuel pressure and injector flow rates are adequate under most driving conditions, significant horsepower and acceleration improvements can be gained. Again, keep in mind that power comes from better handling of the air coming into the engine. It is easy to get more fuel into the engine.

Fuel Pressure

On the stock TPI system, the fuel pressure is regulated at 35 psi with no load on the engine. As the engine is put under a load, the fuel pressure increases in one step, jumping up to between 40 and 45 psi. There are two types of fuel-pressure regulators currently on the market that can significantly improve power.

The first is an adjustable fuel-pressure regulator. Dynamometer tests show that horsepower and acceleration improve with about a 20–25 percent increase in fuel pressure. TPI Specialties' tests on their adjustable pressure regulator installed on a stock 350 ci using dyno headers show an increase of more than 12 hp over the stock fuel pressure at 4,500 rpm.

While at the track with this unit installed you will need to disconnect the oxygen sensor. When the ECM enters the closed-loop mode the oxygen sensor would report the extra enrichment caused by the higher fuel pressure to the ECM. The injector pulse width would then be shortened, leaning the air-fuel ratio, and defeating the benefits of the higher fuel pressure. Disconnecting the oxygen sensor would either keep the ECM from entering closed loop or deliver a neutral air-fuel ratio signal to the ECM if it does.

Micro Dynamics of England builds a rising-rate fuel-pressure regulator. This unit is adjustable and accurately tracks manifold pressure to precisely alter fuel pressure to meet engine demand. Again, street legality is highly questionable, so you might want to save it for the track. The advantage of the rising-rate regulator is good top-end performance without sacrificing a good idle. The United States price for the Micro Dynamics rising-rate regulator is about $300.

Since the early 1990s, a growing number of aftermarket performance parts manufacturers have responded to enthusiasts' cries for more power. This Edelbrock intake is Chevrolet's TBI-equipped 454 trucks, like the SS-454, and is emissions-legal. Edelbrock also produces emissions-legal exhaust headers, TPI manifolds and runners, and even complete fuel injection systems to retrofit onto non-EFI vehicles to improve their performance. *Edelbrock*

Other companies that market 45 psi or adjustable regulators include Digital Fuel Injection and Hypertech (see appendices for addresses).

Changing fuel pressure is like changing a carburetor's float level: Raising the float level a little can increase an engine's performance, and raising it a lot can destroy it. You will need to play around with it a little to determine the fuel pressure that is right for your engine.

So-called Street/Strip modifications always give the best performance improvements when two or more are planned and executed together. Keep in mind that the power from an internal-combustion engine mostly comes from its ability to gulp, ignite, and expand air. Increasing the fuel pressure reaps its maximum benefit when coupled with one of the engine breathing modifications mentioned earlier in this chapter. Again, it is easy to put in more fuel but the power comes from more air, and more air is hard to put in.

Injectors

Different electronic fuel-injection applications use different injectors and different injectors have different flow rates. The TPI 5.7-liter engine used an injector with a flow rate of about 22 lb-hr. The 1989 Ford Thunderbird Supercharged 3.8-liter Super Coupe used an injector that flows 32 lb-hr. At first glance this might seem to be a favorable improvement, yet the end result of putting the Thunderbird injectors in the Corvette might well turn out to be the same as raising the float level too high on a carburetor. Injectors cost $30 or more on the cheap! For a V-8 this means a minimum of $240 to replace the injectors and a more realistic figure would be closer to $500. This investment would show almost no positive gain in performance and may cause drivability problems. Your money would be better spent in camshafts and recurving the distributor.

Required injector flow rates can be calculated using the following formula:

$$lb\text{-}hr = \frac{\text{Maximum hp} \times \text{brake specific fuel consumption}}{\text{Number of injectors}}$$

The brake specific fuel consumption relates to the amount of fuel it takes to create 1 hp. This figure is typically 0.45 on a normally aspirated and 0.55 on a turbocharged or supercharged engine.

Using a horsepower figure of 350 for a 350-ci Corvette engine, the formula would look like this:

$$\frac{350 \times 0.45}{8} = \text{approx. 20 lb-hr}$$

Since the stock injector has a flow rate of 22 lb-hr, it will be more than adequate for modifications up to about 390 hp.

A highly modified turbocharged engine with a maximum horsepower of 500 would look like this:

$$\frac{500 \times 0.55}{8} = \text{approx. 35 lb-hr}$$

An engine like this would require either an aftermarket fuel-injection system or an alternative fuel source when operating under boost. To solve this problem, some performance modifiers have rigged the cold-start injector to energize when the turbo goes into boost. The problem with this is that there is no way to accurately control the amount of additional fuel, and not all applications you might be turbocharging have a cold-start injector.

Micro Dynamics builds an electronically controlled auxiliary fuel injector for situations like this.

Electronic Controls

The area with the greatest potential for cheap performance is the electronics of the injection system. Back in the old days (prior to 1980), any responsible performance specialist would accompany intake, engine, fuel, and exhaust modifications with changes in the timing advance curve as well as adjustments to the vacuum advance and power valve. Today, these adjustments are made by the computer as the car is being driven.

The program in the computer does not determine initial timing; that is determined by the physical position of the distributor or by the crankshaft sensor. But it does determine the amount of ignition advance based on coolant temperature, engine load and rpm. Additionally, the computer determines the amount of enrichment required by each of the above engine-operating parameters, as well as throttle position.

Since the computer plays such a dominant role in the control of two vital areas of engine performance—timing and load enrichment—any significant improvement in engine breathing must be accompanied by an improvement in the computer program. When Chevrolet designed the computer program for the 5.7-liter Tuned Port Injection system, it had performance in mind, but it also had to deal with the EPA. As a result, the program in the stock computer is a compromise between power, economy and emissions. Compromises for the manufacturers are mandated to lean toward minimizing pollution and maximizing economy at the sacrifice of power.

There are three ways of overcoming this sacrifice of power when your car goes to the track: Cheap tricks; performance chips; and performance computers.

Cheap Tricks

There are a lot of racer's tricks that have been rumored about during the last few years. Most of these involve dumping more fuel into the engine and are marginally effective at best. Here are a few tricks that do show an appreciable improvement in performance.

640 Ohm Coolant Sensor

The 640 ohm resistor is probably the cheapest and easiest performance modification. Simply replace the Coolant Temperature Sensor (CTS) with a 640 ohm resistor. To do this, disconnect the CTS harness from the CTS, then install the resistor across the terminals of the harness. The 640 ohms of resistance tells the computer that the engine is not quite warmed up yet, so the computer will allow more advance since it does not expect detonation when the perceived engine temperature is only about 145 to 150 degrees Fahrenheit. The net result is like recurving the distributor, rejetting the carburetor and making power-valve refinements.

If you use this trick, its success will depend on the initial timing being adjusted properly and the quality of fuel you are using. If the car has poor-quality gas in it, the tendency for detonation increases. (Remember, the computer only believes that the engine is running cold; in reality it is at operating temperature.)

Be warned, though, other systems may be affected by the installation of the 640 resistor. For instance, the torque converter clutch may be inhibited from operating with the resistor installed. On the 5.7-liter engine and others, torque converter lock-up is inhibited at temperatures below 150 degrees Fahrenheit. While this can improve performance by eliminating the bog inherent as the converter engages, prolonged operation without lock-up can cause the internal temperatures of the transmission to increase, resulting in potential damage.

On some applications, such as the 2.0-liter Corsica and Beretta, there is no back-up coolant fan switch. As a result, the coolant sensor input is the only way the computer has of knowing when the radiator fan needs to be switched on. Leaving the 640 resistor installed when you leave the track can result in severe engine overheating and damage.

This is a cheap but effective modification—about 19 cents will take care of two cars.

Power Thermostat

A double benefit for a power increase can come from installing a cold thermostat. A 160-degree Fahrenheit thermostat will have a similar effect on the computer as the 640 ohm resistor, allowing additional advance and a richer mixture with the oxygen sensor disabled.

There is an added advantage that the resistor trick did not offer: The lower engine temperature results in less heat being transferred to the intake manifold. The lower intake manifold temperature causes the air in the intake to be heated less, so the air remains denser and therefore capable of delivering more power. To a limited extent you get some of the effect of a turbocharger without the expense.

The lower air charge temperature also serves to lower combustion chamber temperatures. This reduces detonation, which would cause the ESC (Electronic Spark Control) system to retard the timing, thereby reducing the power. The net result is more timing advance and therefore more power.

You might be wondering at this point why the manufacturers would not use a 160-degree thermostat as standard equipment if lower combustion temperatures would result. After all, lower combustion temperatures would result in a lower production of oxides of nitrogen (NOx). True as that may be, lower combustion temperatures result in a higher level of carbon monoxide (CO) and hydrocarbons (HC) being emitted. Again power is sacrificed to pollution. Keep this in mind, however, if you chose to install a cold thermostat. This simple operation may be in violation of smog laws, and may result in emission levels that are high enough to fail your smog test.

Power Fan Switch

Along the same line of thinking and in conjunction with the colder thermostat, there is the low-temperature fan switch. The stock Chevrolet radiator fan switch will turn the fan on at about 220-degrees Fahrenheit and off around 180-degrees Fahrenheit. Intake manifold air density can be increased by installing a switch that turns the fan on at around 176-degrees Fahrenheit and off at around 166-degrees Fahrenheit.

On some applications, the cooling fan is controlled by the ECM. The

Here's an aftermarket clone of Chevy's TPI system for a big-block 454. Chevrolet has used similar systems on "development mules" that produce as much as 700 hp or more!

only way to lower the fan-cycling temperatures on these is to install one of the performance PROMs discussed later in this chapter.

MAP Sensor Delay Valve

When the engine comes under a load, as when the car accelerates, manifold pressure increases (vacuum drops). The increasing manifold pressure signal from the MAP causes the ECM to retard the timing to lower combustion temperatures, reducing detonation and the possibility of NOx.

Installing a short-delay vacuum delay device in the vacuum hose between the manifold and the MAP sensor can slow the increase in pressure to the MAP and allow the timing to advance faster. The down side of this is that the MAP also performs the power-valve function for the ECM; if there is too much of a delay, the air-fuel ratio will not enrich fast enough and the car will stumble or hesitate, defeating what you have tried to accomplish.

Oddly enough, a good source for these delay valves is your local Ford dealer. Back in the late 1970s and early 1980s these vacuum delays were a vital element in Ford's emission control program. A vacuum

delay of 1 second or so is about right. You may have to try several; for your car, slightly more or slightly less may do the trick. Basically you want the delay to last no longer than the time it takes to move the throttle from the closed position to the wide-open throttle position.

Sometimes after going through the expense of trying several vacuum delays you will find that none of them significantly increases power. It could be that to achieve what you are trying to accomplish, all that is needed is a restriction in the MAP sensor vacuum hose. Play around a little with vacuum line and electrical butt connectors.

Whether a butt connector or a vacuum delay is used, the beauty of them is that they can be installed and removed easily at the track.

Of course, the MAP sensor is not used on most TBI or TPI cars, so this trick is not applicable to those engines.

Performance Chips

Imagine a day in the future when you can jump in your car, start the engine, back out of the driveway, and motor off quietly and smoothly to the racetrack. Once in the pits you reach out to the instrument panel and throw a switch. The tame little kitten becomes a flame-breathing lion: Cam timing, lift, duration, and overlap have changed, the mixture has enriched to a performance level, and the distributor has been recurved. This is a picture of the future, all of which but the cam changes are here today.

Several companies make aftermarket computer calibration chips (or PROMs, which stands for Programmable Read Only Memory) for Chevrolets. One of the most prominent is Hypertech (see appendices for address).

These performance chips alter the event points of the ECM. They may also do the following:

• Alter the temperature at which closed loop is entered (if at all)
• Decrease the amount of manifold pressure required to enter enrichment mode
• Decrease the throttle position voltage needed to enter enrichment mode
• Accelerate the timing curve based on MAP and rpm increases
• Recalibrate the activation of peripheral emission control devices such as EGR, lock-up converters, canister purge, and air pump switching

Where Chevrolet leaned toward emissions and economy in the aforementioned compromise, Hypertech and its competitors leaned toward performance.

These chips are probably the fastest and easiest route to a performance upgrade available for late-model Chevrolet fuel-injected cars. To install an aftermarket performance chip, remove the ECM from the vehicle, then remove the screws that hold the access cover on the side of the ECM. Remove the access cover and locate the CALPAK.

Note: Be careful of static discharge. Even a small zap can destroy the ECM. Before removing the CALPAK, discharge any static electricity that may have built up in your body during the day. Radio Shack and similar electronics stores sell grounding straps that fit around the wrist and attach to a good ground. These wrist straps ground your body and prevent the build-up of static electricity. At an investment of less than $10, purchasing one of these is a good safety precaution.

After taking the appropriate precautions against static, carefully remove the CALPAK. In some applications this will require extraordinary care as the chip is mounted in a DIP (dual inline package) socket, and when they are removed the pins are fully exposed to potential bending and other damage. On later models, the chip is mounted in a holder, which reduces the potential of damaging it either physically or through static electric discharge.

Once the old chip is removed, inspect the socket on the mother board for damage and install the new chip, being careful not to damage it.

These aftermarket chips come in three performance grades; Street stock, Performance Modified and Competition. These chips are not California legal, however; they fall into Category III of the BAR list of aftermarket parts. There is one company, Street Legal Performance, that does market a California-legal performance/economy PROM.

Performance Computers

For total performance applications, companies like MicroDynamics, represented in the United States by Veloce Distributing (see appendices for address), offer custom-designed performance computers. These computers are offered as either a replacement for the ECM or as a piggyback add-on to the current ECM. These units offer very little over the performance chip until major engine modifications such as turbocharging, high-compression pistons, and intake modifications demand it. On a standard street car, where the owner is looking more for power improvements than for the ability to run the quarter mile at warp speed, the performance chips are adequate.

Modifying OBD-2 Engines

Changing regulations are making performance modifications tougher and tougher all the time. With the advent of OBD-2, there is virtually no room for performance modification. Basically, if there is intent to keep the vehicle licensed for use on the street, no modifications are permitted without specific approval by the California Air Resources Board (CARB). Gaining approval from CARB for a performance product is difficult, time consuming, and expensive. As a result, there are few approved products, and I expect it to remain that way. At least part of the lack of motivation is the fact that there is no longer a pervasive interest in performance cars.

C
O
N
T
E
N
T
S

**CATEGORY 1
REPLACEMENT
PARTS** 154

**CATEGORY 2
REPLACEMENT
PARTS** 154

**CATEGORY 3
REPLACEMENT
PARTS** 155

**ARB-APPROVED
PARTS
AND SUPPLIES** 155

**ELECTRONIC
CONTROL UNITS
(ECU)** 155

**EXHAUST
COMPONENTS** 155

**INTAKE
COMPONENTS** 155

**FUEL INJECTION
SYSTEMS** 155

**THROTTLE BODY
INJECTION UNITS** 155

SUPERCHARGERS 155

CHAPTER TWELVE

ENGINE MODIFICATIONS AND THE LAW

Unlike most scientific regimens, the law has subtle twists, turns, and traps. This chapter is not intended to be legal advice; it is intended to point out some of the issues involved in the high-performance modification of late-model cars. The bulk of the information that follows came from California's Bureau of Auto Repair (BAR) and the California Air Research Board (ARB). The laws on the books in many states are similar to those of California. However, California has the longest record of stringent enforcement, which is why I am using their information.

There are three categories of replacement parts recognized by the ARB. Each category is described in the following paragraphs.

Category 1 Replacement Parts

Category 1 items are not considered by the BAR or ARB to be of any concern as long as the required emission controls are not tampered with. Examples include:
• PCV air bleeds
• Air cleaner modification
• Air conditioner cut-out systems
• Anti-theft systems
• Blowby oil separators and filters
• Electronic ignition systems retrofitted to vehicles originally fitted with point-condenser systems as long as the original advance controls are maintained
• Engine shut-off systems
• Ignition bridges and coil modifications
• Throttle lock-out systems
• Intercoolers for OEM turbochargers
• Under-carburetor screens
• Vapor, steam or water injectors

Category 2 Replacement Parts

Category 2 addresses allowable replacement parts:
• Headers on non-catalyst cars
• Heat stoves for allowed headers
• Intake manifolds for non-EGR vehicles must allow for the installation and proper functioning of the OEM emission controls
• Approved aftermarket catalytic converters
• Carburetors marketed as emission replacement
• Replacement fuel fill pipe restrictors
• Replacement gas caps

You can see from this list that for catalyst-equipped fuel-injected cars there are no performance replacements that are allowable without type approval from the ARB. Today's production cars are EPA-inspected as an integrated system; disturbing even the most minute portion of the emission control package would constitute a violation.

Category 3 Replacement Parts

Category 3 parts must have verification of acceptability. If you are replacing a part in Category 3, ask for and retain a copy of the verification of acceptability for that product. It may prove handy later on, even if you live in an area that is not currently strictly controlled. Category 3 includes:

- Carburetor conversions
- Carburetors that replace OEM fuel injection
- EGR system modifications
- Replacement PROMs (computer chips)
- Electronic ignition enhancements for computerized vehicles
- Exhaust headers for catalyst vehicles
- Fuel-injection systems that replace OEM carburetors
- Superchargers
- Turbochargers

ARB-Approved Parts and Supplies

What follows is a list of the ARB-approved equipment and suppliers as of January 1, 1990:

Electronic Control Units (ECU)

Street Legal Performance Engineering, Inc.

Engine calibration software EPROM for 1985–1988 Camaros with 5.0- and 5.7-liter Tuned Port Injection engines

Exhaust Components

Gale Banks Engineering

Exhaust crossover pipes for 1982–1990 5.0- and 5.7-liter Camaros with Crossfire Injection or Tuned Port Injection

Street Legal Performance Engineering, Inc.

Tubular exhaust manifolds for 1985–1990 5.0- and 5.7-liter Camaros with Tuned Port Injection

Intake Components

Street Legal Performance Engineering, Inc.

Intake manifold runners for 1985–1990 5.0- and 5.7-liter Camaros with Tuned Port Injection

Fuel Injection Systems

Air Sensors, Inc.

Electronic Fuel Injection for 1986 and earlier Chevrolet 454-ci heavy-duty gasoline engines (trucks)

Throttle Body Injection Units

Holley

Throttle Body Injection for 1982–1986 2.5-liter and 1983-1986 2.0-liter engines

Superchargers

K.F. Industries, Inc.

Max-25 Supercharger for 1986–1988 Chevrolet light-duty trucks powered by a 2.8-liter GMC V-6 fuel-injected engine

As you can see by the briefness of this list, there is not a lot of flexibility in California for legal modification of a late-model fuel-injected car. Since the laws in California, to some extent, reflect a federal mandate, it can be assumed that the regulations in your jurisdiction may become subject to the same requirements as California.

The part that might prove both financially and emotionally frustrating is that laws and regulations are subject to change, sometimes resulting in a retroactive ban on some performance modifications. If you intend to use your car on the street for an indefinite period in the future, then it would be worthwhile to consider the cost and time requirements of returning the system to factory stock if the need should ever arise.

Cars not intended for street operation are still wide-open territory in most jurisdictions. There are, even for full-fledged race cars, noise ordinances to consider, however.

CONTENTS

BOLT-ON
SYSTEMS   156
HOLLEY PRO-
JECTION SYSTEM   157
AIR SENSORS
FIRST SYSTEM   157
DIGITAL FUEL
INJECTION
SYSTEM   157
AIR SENSORS
MULTI-POINT
SYSTEM   157
SELECTING AN
AFTERMARKET
FUEL-INJECTION
SYSTEM   157

CHAPTER THIRTEEN

AFTERMARKET FUEL-INJECTION SYSTEMS

Only a few years ago, the idea of buying an electronic fuel-injection system over the counter brought forth images of young men with uncombed hair wearing broken horn-rimmed glasses and an oxford shirt with 23 pens in a pocket protector. Today, we find systems requiring installation skill levels ranging all the way from the occasional back-yarder to those of the most expert professionals.

Bolt-On Systems

The simplest systems to install are the direct bolt-on Throttle Body Injection-style systems. These systems can have as few as four components to install—the throttle body assembly, the fuel pump, the coolant temperature sensor, and the computer.

Throttle Body Systems

Since these systems have been designed with either performance or economy in mind, they are generally formatted as a twin-injector two-barrel system. Using two low-pressure side-feed injectors similar to the injectors used by GM on their late-model applications, the throttle body installation is as simple as installing a carburetor. The TBI unit includes the fuel-pressure regulator, which, unlike those on the pollution-controlled injection systems, is adjustable. The fuel pressure is pre-set, but adjustment can be made when necessary on a given installation. The TPS (throttle position sensor) comes factory adjusted, but provision is made for further adjustment if the need should arise.

Cold-start fast idle is achieved through a one-step fast idle solenoid kicking the throttle open during warm-up.

Fuel Pump

For ease of installation these systems use an inline, chassis-mounted electric fuel pump. When this pump is installed, it should be mounted as close to the fuel tank as possible. The vane pump does an excellent job of pushing fuel but a lousy job of pulling fuel. As the pump is mounted farther away from the tank, the chances of vapor lock on hot days with a low tank increases dramatically.

Temperature Sensor

A temperature sensor is used to detect the coolant temperature. This sensor screws into any available threaded hole in the water jacket of the intake manifold.

Electronic Control Unit

The electronic control unit, or ECU, is connected by way of a wiring harness to switched ignition voltage, the TPS, the coolant sensor, a tach signal from the coil, the crank terminal of the starter, the injectors, and a good ground.

Most of the ECUs in this category of after-market injection systems are adjustable through the use of potentiometers. A typical ECU would have five adjustment pots.

- Choke—adjusted to provide sufficient extra fuel for cold starting and warm-up operation.
- Accelerator pump—on the emission injection systems, this adjustment is handled by ECM calculations based on inputs from the TPS and the MAP sensor. Since many of these systems do not use a MAP sensor, this adjustment is used along with the TPS input to trim the air-fuel ratio required for initial acceleration.
- Idle—this may well be the most critical adjustment of any of them. The rest of the adjustments can be made for best performance or smooth running. Adjusting the idle fuel to the point where the engine runs the smoothest would cause you to fail the emission test in most every state in the United States that does emission testing.

    The target market for the injection systems in this category are small-block and big-block V-8s of the 1960s, 1970s, and early 1980s. These cars tend to idle their smoothest at between 4- and 6- percent carbon monoxide. These levels would simply fail most emission tests. The best way to make this adjustment, therefore, is by means of the old lean-drop method of the carbureted cars of the 1970s. Adjust the pot for the smoothest idle at about 25 rpm above the desired idle speed, then lean it out until the rpm drops 25 down to the desired level.
- Midrange—this adjustment could most closely be compared to the main circuit of a carburetor. This pot should be adjusted to the setting that provides for the best and

smoothest cruise operation.

If you prefer to keep it simple, Holley manufactures a line of systems known as the Pro-Jection Fuel Injection System. They are suitable for applications of all the American manufacturers from displacements of 301 to 400 ci.

This system features very few components to be installed and very wide coverage from a limited number of part numbers. Stressing simplicity, these systems could easily be installed by anyone possessing enough skill to replace a carburetor.

Performance Systems

For those who are looking at big-block applications, or for small-blocks that have been modified a little, there are a couple of four-barrel throttle body applications available.

Holley Pro-Jection System

The Holley four-barrel system is known as the Pro-Jection 4 system. Other than the extra two barrels in the throttle assembly there really is not a lot of difference between this system and the one already described.

This system is rated by Holley at 900 cfm and with sufficient fuel flow to meet the needs of a 600 hp engine.

Air Sensors FIRST System

A company from Seattle, called Air Sensors, markets a sequentially-fired four-barrel throttle body system. Unlike the Holley, this is a high-pressure system where the ECU opens the injectors one at a time.

Another difference is that the Air Sensors system uses a hot-wire Mass Air Flow sensor of its own design, which is similar in concept to the Bosch MAF used on the Chevy Tuned Port Injection engines. The addition of the MAF provides more precise monitoring of the air mass moving into the engine and therefore more precise control over the air-fuel ratio. In fact, the Smog Book for California BAR lists this system as an approved Category 3 conversion for the 1986 and earlier light-duty trucks equipped with a 454-ci engine.

Although a little more complicated to install than the Holley system, installation of the Air Sensors FIRST Throttle Body system would require the skills of an experienced performance modifier.

Air Sensors rates this system at

1,000 cfm with capability of a sustained 375 horsepower. At first comparison, it may look as though the Holley system has abilities beyond the Air Sensors system. However, Air Sensors is quoting specifications relating to support for sustained horsepower, where Holley is referring to engines with brief peaks of horsepower up to 600.

Several companies now market fuel-injection systems that range from clones of the Tuned Port Injection system to Big Block Tunnel Port systems.

Digital Fuel Injection System

Digital Fuel Injection (DFI) (see appendices for address) markets a Tunnel Port fuel injection system that offers either a 650 or 1,000 cfm throttle body with injectors rated up to 500 hp. These are speed-density systems (they use a MAP sensor) using a factory-programmed computer. For the real enthusiast, DFI offers a software package and serial interface cable so the end user can modify the program.

DFI offers systems rated up to 2,000 cfm and 700 hp. To a large extent this is a user-friendly, bolt-on system.

Air Sensors Multi-Point System

The challenge of the supercar category has also been met by Air Sensors, which offers a TPI-clone system: A high-performance bolt-on multi-point injection system with breathing capabilities up to 450 hp. This system uses Air Sensors' own MAP and a user-adjustable digital ECU.

Another Air Sensors offering is the Multi-Point injection system. This one is not to be installed by the faint of heart. But for those who enjoy a challenge and possess either the skills or the money necessary to modify the existing or aftermarket manifold to accept the injectors, this is an exciting system.

Selecting an Aftermarket Fuel-Injection System

There are several important things to consider when you are selecting an aftermarket injection system: your skill level; why you are considering an injection system; how much money you have to spend; and how much vehicle downtime you can afford while making the conversion.

Your Skill Level

Be honest. Overestimating your

abilities can be a costly mistake. The skill required to install these systems ranges all the way from the Saturday mechanic with only a single free afternoon, to the level of master mechanic/master machinist.

Why You Are Considering Fuel Injection

Frankly, some of the systems on the market today compared with the price of many new replacement carburetors make throttle body injection a viable substitute. These simple throttle body systems are also reasonable substitutes for fuel economy. While the savings are not on a par with the legendary 200 mpg carburetors, many are well documented to save between 5 and 10 percent on fuel.

For those who are a little more performance oriented with a little more skill, time and/or engine, there are relatively simple performance throttle body systems and PFI clones available.

How Much Money You Have to Spend

Money is always a consideration, and as you are planning your upgrade to fuel-injection, you should bear in mind that the fuel-injection kit may be among the least of your expenses in this changeover.

The following potential pitfalls may or may not be of concern to you, depending on the type of injection system you buy and the condition of the engine you are installing it on.
- Compression and vacuum condition of the engine. This consideration is especially important on MAP (speed/density) systems
- Type and condition of the camshaft
- Intake manifold modifications
- Exhaust modifications required if the system uses an oxygen sensor
- Customer support for problems and replacement parts

How Much Vehicle Downtime You Can Afford

Some of these systems require only a few hours of soda pop-sucking wrench time, while others require weeks of planning, estimating, and trips to the machine shop. If the car you are installing the system on is your only means of transportation, you will probably not be interested in the more complicated or sophisticated systems.

CONTENTS

**OBD-2
DIAGNOSTICS—
THE CONCEPT   159**

**SO WHAT
CAN I DO?   159**

**SYSTEM
ANALYSIS CHART 163**

**OBD-2 CODE LIST:
POWERTRAIN
DIAGNOSTIC
TROUBLE CODES 164**

CHAPTER FOURTEEN

ON-BOARD DIAGNOSTIC SYSTEM GENERATION 2

Since the first printing of this book, there have been two major changes in Chevrolet fuel injection: The first is simplicity—today's systems have fewer components, because engineers have managed to eliminate peripheral emission control devices such as EGR valves and air pumps; the second change is a little more complex—on-board diagnostics. Since 1994, there has been a progressive industry push and government mandate for a new, universal diagnostic system. The new diagnostic system is called OBD-2, which stands for "On-Board Diagnostic System Generation 2." This system, or maybe it should be referred to as a concept, was touted to the industry as a standardized language and diagnostic method. Each of the manufacturers is to adhere to very tightly specified guidelines with respect to communication protocols and serial data. At this point, you are probably a bit glassy-eyed, wondering what the heck I am talking about.

In the early 1980s, General Motors designed a diagnostic system consisting of a set of diagnostic codes and a "stream" of information that showed what the computer was seeing, hearing, feeling, smelling, thinking, and doing. This information is known as serial data. At the time this method was developed, it was far beyond what other manufacturers were doing with their diagnostics. In 1984, Ford introduced the fourth generation of its Electronic Engine Control system, the EEC-IV, into mass production. Along with this came a diagnostic system utilizing a sophisticated set of codes.

Because of ever-stricter laws governing emissions output, manufacturers must now engineer powertrains as systems, in order to meet the federal requirements. Here, a 1997 Corvette's 5.7-liter LS1 engine, PCM-controlled four-speed automatic transmission and rear axle assembly are tested on a dynamometer. *Chevrolet Motor Division*

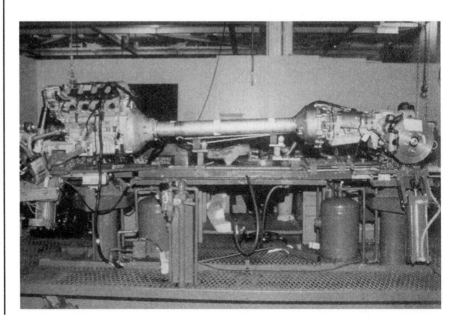

When the EEC-IV system delivered a code to the technician, it, in effect, meant that the computer had analyzed the information the GM system would have presented in the data stream and found that one of the pieces of data indicated a problem. Where the GM system left it up to the technician to determine if a presented parameter was correct, the Ford system did that analysis itself. Chrysler then went one better and provided the technician with a method of testing the actuators. The Actuator Test Mode (ATM) allowed the Chrysler technician to check the computer's ability to control the ignition system, the injectors, the alternator, and virtually every other system under the hood.

OBD-2, at least on the surface, is styled very much along the lines of the GM diagnostic system. The system has a potentially very large number of codes that the manufacturer may choose to include in its diagnostics. Additionally, again like GM, the OBD-2 system incorporates a thorough serial data stream. Therefore, the technician or enthusiast that is comfortable with GM diagnostics will be able to make a relatively easy transition.

OBD-2 Diagnostics— The Concept

With the old OBD system, GM had three different kinds of codes. "Hard codes" indicated that the system had a defect that was being detected at that moment. When there was a hard code, the Check Engine light would be illuminated by the Engine Control Module (ECM). A "soft code" was the result of a hard code that had somehow corrected itself, or at least was no longer being detected by the ECM. The third type of code was Code 12, which indicated that the ECM was not receiving a pulse from the crankshaft rotation sensing device. Code 12 was always received when attempting to pull codes because the engine would not be running when the codes were being pulled. With OBD-2, an additional code type is added to indicate

or confirm to the Powertrain Control Module (PCM) that the item being monitored is actually working.

My first experience with OBD-2 was on a Lumina. I demonstrated the power of the distributorless ignition system for one of my classes by disconnecting the two plug wires on one of the ignition coils. This, of course, caused the engine to idle unevenly. The computer noticed changes in the acceleration rate between the various cylinders and set a Code PO301. This code identifies a misfire on cylinder number 1. We all pondered the ability of the computer to recognize this problem with some amazement. We then shut the engine off, reconnected the disconnected plug wires, and restarted the engine. Much to my chagrin, the Service Engine Soon light (the OBD-2 term is "MIL", for Malfunction Indicator Light) remained illuminated. I feared that I might have had to take this car back to the rental company with the MIL glowing. The next morning, the light went out as the engine approached operating temperature. The PCM needed to see a smooth running engine that remained smooth throughout a warm-up cycle in order to cancel the MIL for a code PO301.

This need for the computer to see a response has been reported to cause some frustration in some quarters. Imagine the above situation taking place in Fairbanks, Alaska, in the middle of January. Imagine that I lived about 1 mile and about 3 minutes from where I worked. All of this is true, by the way. From the time I left home until the time I got to work, not enough time would pass to allow the engine to warm up. Also, the temperature was 43 degrees below 0 Fahrenheit. At that temperature, it may take 15 to 30 minutes for the engine to approach operating temperatures. Where these extremes of climate exist, a code could cause the MIL light to remain illuminated since the PCM may never see a complete warm-up cycle.

These potentially troublesome codes are called readiness codes.

They are set when the PCM has not noticed a sensor or actuator circuit working properly or working at all. Readiness codes do not apply to all circuits. When the battery is disconnected, there may be several of these codes generated that will not clear until the computer has seen a specific set of circumstances and signals occur. Even after a successful repair is made, the MIL may remain on until a warm-up cycle is completed. This is inconvenient for the do-it-yourselfer, since he or she may not know for sure that the attempted repair was successful.

The basic diagnostic tool for OBD-1 systems was—and remains—the voltmeter. With OBD-2, the basic diagnostic tool is the scanner. Without this rather expensive piece of equipment, you are almost dead in the water. A professional scan tool will set you back about $2,000. Even for the "amateur" unit, the price can easily approach $1,000. The big question the reader is probably asking right now is: "Do I need a scanner to diagnose my own car?" Unfortunately, the answer is yes. There are a couple of manufacturers of aftermarket test equipment that produce small, inexpensive testers for OBD-1. I look for these manufacturers to eventually produce a relatively inexpensive OBD-2 scan tool within a few years. Doing this, however, will be quite a bit more complex than it was for OBD-1. The price will most assuredly be greater than the $50 to $80 typical for the low-end OBD-1 tools.

So What Can I Do?

Troubleshooting OBD-2 without a scanner can be tedious and almost futile at times. Without a scanner, you cannot even get trouble codes. So there you are: a MIL is staring you in the face; maybe there is a driveability problem, maybe there isn't. How do you proceed? Basically, what is required is the same type of troubleshooting that was required on the Bosch electronic fuel injection systems of the early 1960s.

Consider the Symptoms

There are symptoms that are associated with specific components and with specific defects in those components. There is a temptation to use trouble codes early in the diagnostic procedure. This temptation should be fought for as long as possible. Real-world experience with these systems indicates that the system defects that cause a driveability problem are seldom system defects that also set trouble codes.

Low Power

There is a common misconception that low power is the result of a lack of fuel. There is a great deal of truth to this when talking about diesels. There is little truth to the statement when talking about spark-ignition engines. Low power in a spark-ignition engine is usually caused by a lack of air volume entering the engine. Sometimes the low power is caused by inadequate heat.

Lack of Air

Air, or lack of it, is the most common cause of poor power. Restrictions to airflow can occur on the intake side or on the exhaust side of the engine. The intake side can be obvious, dirty air filters are often responsible for poor power. Less obvious are things that restrict the movement of the throttle pedal or the throttle plate.

Spark

Poor spark delivery can also reduce power. As stated above power comes from air; but to be more specific, power comes from expanding the air. If there is an adequate amount of air being drawn into each cylinder and if there is an approximately appropriate amount of fuel being carried into the cylinders by the air, then the power will be created by ignited the air/fuel charge with a spark. The absence of this spark, or poor quality spark can prevent the proper ignition of the air/fuel charge and therefore reduce power.

Compression

One of the primary jobs of compression is to raise the temperature of the air/fuel charge in the combustion chamber. Low compression reduces the temperature of the charge at the point in the compression stroke where the spark plug fires. If a cylinder has low compression, the quality of the combustion process will be reduced. Low combustion quality will result in high emissions, of course, but also will result in reduced power from that cylinder. Low compression can cause low power, but the most commonly described symptom resulting from low compression is rough engine operation, especially at idle.

Ignition Timing

Ignition timing is critical for proper production of power from the engine. Many of today's engines feature distributorless ignition systems that do not make allowances for changing or adjusting the ignition timing. Firing the spark plug too soon will ignite the air/fuel charge before the compression has increased the temperature to the point where the spark will yield maximum power from the charge. Additionally the charge, once ignited, will attempt to drive the piston back down the cylinder before it passes over top dead center. In other words, the expanding gases resulting from the ignition of the air/fuel charge tries to make the engine run backwards. Again, power output of the engine is reduced and the emissions go through the roof.

Summary

In my experience as an automotive technician, and as a fuel injection specialist in particular, one of the most common complaints I have heard is of low power. Power is a relative thing. It seems that another car always has more power, that your car had more power, or that your car should have more power.

The fuel injection computer and the system associated with it are primarily designed to deliver fuel and to control the ignition timing. Power production is most closely associated with the ability of the engine to breathe. Although fuel and timing related problems may cause a loss of power, the place to start a troubleshooting routine is by looking for restrictions in the air-induction system or the exhaust system.

Runs Rich

I cannot remember a customer ever coming to me and saying, "My car is running rich"—that is has an imbalance in the air/fuel ratio. Instead, they speak of the problems symptomatically. Typical symptoms when the engine is running rich include brown or black smoke, a raw fuel odor, a rotten egg smell, or poor fuel economy.

Although the most common user complaint is poor fuel economy, there are much more serious consequences if an over-rich condition is left unrepaired, and two are quite serious: An over-heated catalytic converter; and expensive internal engine damage.

First, an engine that is running rich produces high levels of carbon monoxide (CO) and hydrocarbons (HC). When these enter the catalytic converter it tries very hard to convert these gases to carbon dioxide (CO_2) and water. This process causes heat to be generated. The more CO the more heat; the more HC the more heat. Soon the temperature of the catalytic converter will be high enough to begin melting the substrate. Melting of the substrate reduces the cross-sectional area of the converter and this causes temperatures to increase even more. Soon the catalytic converter is ruined, the exhaust system becomes restricted and there is a noticeable loss in power.

Second, when the fuel flow is excessive, the gasoline—an excellent "solvent"—tends to wash the oil off of the cylinder walls. This "wash down" reduces the lubrication of the cylinder walls and that can result in engine deterioration. Engine overhauls that are instigated by wear of the cylinder walls are always expensive.

A rich running condition is, indeed, a problem that is likely to be caused by the fuel injection system. Either the engine is getting too much fuel or not enough air. Check the air filter for restrictions caused by dirt, oil, or other contaminants. Check the exhaust system for restrictions, as well. If the engine is unable to remove the spent exhaust gases from the cylinders, then it will be unable to draw in fresh air for combustion.

If there is no evidence of restriction to airflow, then it is time to get into the system that controls fuel flow. The first thing to check when you suspect the fuel injection system to be the cause of a driveability problem is the fuel pressure.

High fuel pressure can cause the engine to run rich, because the fuel injection computer has been programmed to assume that the fuel pressure is a specific amount. Flow rate is directly linked to the pressure of the fuel: Low fuel pressure can cause an insufficient supply of fuel, while high fuel pressure will cause an excessive flow of fuel. Causes of high fuel pressure are restrictions in the fuel system return line to the tank or a defective fuel pressure regulator.

Runs Lean

Lean running engines are usually observed by the operator or owner as a stumble, a hesitation, or a bog. While fuel does not determine the actual "power" an engine produces— the expanding gases do—an insufficient amount of fuel reduces the amount of heat generated during combustion. Less heat means less expansion, which means less power. With many ignition systems—especially older ones—the coil may not have enough energy to fire the spark plugs in lean running cylinders. In fact, this can cause a misfire that can cause HC (hydrocarbon) emissions to elevate.

When looking for the causes of low fuel pressure, remember that the fuel pump is a device designed to deliver a *volume* of fuel to the injection

system—pumps do not create *pressure*, they only deliver a volume (restrictions to the flow create pressure). Therefore, low fuel pressure is almost always a low volume problem. The most notorious restriction is a dirty fuel filter. Low pressure can also be caused by kinks in the fuel supply (inbound to the fuel injection system) lines, a bad fuel pressure regulator, a worn fuel pump or a combination of problems. Of these three, the least likely to result in low pressure is a defective fuel pump. Typically when a fuel pump goes bad it fails completely.

During the mid-1980s restricted fuel injectors were very common and were primarily caused by sediments and carbon building up on the external tip of the injector. Reformulation of gasoline and changes in injector and intake manifold design have greatly reduced this problem, though.

Hesitation

Next to fuel economy problems, the most common driveability complaint is probably hesitations. The term "hesitation" is used to describe the sensation that occurs when you move your foot toward the floor and the engine says, "You want me to do what?" before it accelerates.

Hesitation is caused by an insufficient supply of one of the three basic ingredients needed to get power from an engine: Air, fuel, or spark. I have had customers complain of a hesitation when the real problem was a bind in the movement of the throttle linkage. Technically, they are right in their description, because the bound linkage interferes with airflow. A weak spark can affect combustion quality, especially when the air/fuel ratio is incorrect. Upon initial acceleration, the throttle has just been opened and there is a large increase in the mass of air entering the combustion chambers. This causes at least a momentary lean condition. A lean cylinder charge burns less thoroughly than a normal or rich cylinder charge; the spark plug sparks, but not with enough energy to completely burn the air/fuel mixture in the com-

bustion chamber. The result is a momentary loss of power that we call a hesitation. Even if the air-fuel ratio is correct, low power from the ignition system can cause a hesitation.

Stumble

The terms "hesitation" and "stumble" are often used interchangeably. That really is okay, since the causes of each of these problems are pretty much the same. Technically, a stumble occurs when the engine is unable to deliver enough power to provide for the change in engine demand requested by the driver. Causes of stumbles, like hesitations, include an interruption in airflow, inadequate fuel delivery, or a weak spark.

Fails emission test

At this writing, most jurisdictions in the United States test only three gases: Carbon monoxide (CO); hydrocarbons (HC); and carbon dioxide (CO_2). Carbon monoxide is tested to ensure that the air/fuel ratio is correct. Hydrocarbons are tested to verify the quality of combustion, and carbon dioxide is tested to make sure the exhaust system is in good condition. Although problems with the fuel-delivery system can cause any of the three to be incorrect, it is far more likely that the problem comes from malfunctions in the air induction, exhaust, or ignition systems.

Stalling

This may just be the most annoying of all problems to most drivers, and certainly one of the most dangerous. Stalling is caused by the loss or sudden degradation of ignition spark, an interruption of fuel flow, sudden excessive fuel flow, or a sudden blockage to the airflow. Although there are a couple of simple things that can cause stalling, the most common causes are the high-tech items.

The first low-tech thing to check would be the fuel level. When the vehicle accelerates, the fuel will often slosh away from the fuel pickup, leaving the engine starved for fuel long enough to stall. On older, carbureted applications this was not a problem because the carburetor stored fuel in the fuel bowl. On acceleration the fuel might have sloshed away from the fuel pick-up, but it would have sloshed back again before the fuel bowl went dry. You do not have the luxury of a fuel bowl in a fuel injected engine, so any loss of fuel volume through the pump will cause an instant loss in fuel pressure and therefore an immediate decrease in the amount of fuel being delivered to the engine.

Another low-tech cause of stalling would be a short in either the primary or secondary sides of the ignition system wiring harness. Look for breaks in the wires, and especially for damage to the insulation of either the primary or secondary wires to the ignition coil. These inspections are tedious but essential, because even the smallest crack in the insulation can allow the ignition system to ground and cause the engine to stall.

High idle

High idle speed problems can *never* be caused by a problem with fuel delivery. The speed of a gasoline spark ignition engine is determined almost entirely by the mass of air entering the engine. If you open the throttle, you increase the air mass; close the throttle and you decrease the air mass. So, begin by checking the throttle linkage adjustment. Make sure that the linkage is not adjusted so tightly that it partially holds the throttle open. Make sure that the linkage, or cable, does not bind. Also look for sources of extra air entering the engine. A common place for this extra air to leak into the engine is where the throttle assembly is mounted to the intake manifold.

Like everything else talked about so far, the computer controls the idle speed of the engine. After the basic checks referred to above have been completed, there are some computer-related items to test. Those procedures will be described later.

Engine will not idle

Again, engine speed is determined by the air mass entering the engine. If the engine does not idle, but runs at all other speeds, it is usually an indication that there is an insufficient mass of air entering the engine when the throttle is closed. There are three major possibilities here: First, the throttle stop may not be properly adjusted; second, the throttle bore may be coked; and third, the idle control device may be inoperative.

There is an adjustment, called Minimum Air, that is an essential adjustment for the proper operation and, in particular, proper idling of a fuel injected engine. The Minimum Air adjustment determines the amount of air allowed to pass across the throttle plates when they are closed. Incorrectly adjusted Minimum Air can result in tip-in hesitation, stumbling, and stalling on deceleration.

Difficult to start

For most people, one of the most noticeable advantages of electronically fuel-injected engines is the ease with which they start. If the engine runs good once started, it is unlikely that the problem is related to fuel delivery or the ignition system. The most likely cause of difficulty in starting is a problem in the starter motor system. The starter motor requires 100–300 amps (depending on engine size) to crank the engine at sufficient speed for it to start. This equates to about 1,000 to 3,000 watts of power at minimum cranking voltage. That, in turn, translates to about 1.34 to 4.02 horsepower. The bigger the engine, the more horsepower it takes to crank the engine over at proper speed. Resistance in an electrical circuit reduces current flow and therefore reduces the amount of power available to crank the engine over. The result will be slow cranking speed. Slow cranking speed will mean that the engine will be creating less compression in the cylinders. Less compression means poor com-

bustion efficiency. The bottom line is that the ignition system has to work extra hard to ignite the air/fuel charge. A tune up will provide new ignition components and therefore might make the engine start easier, but it will not cure the problem.

Beginning with the battery, test all the components of the starting system. To test the battery, disconnect the primary side of the ignition coil, connect a voltmeter across the battery terminals, and observe the open circuit voltage. Then crank the engine for 15 seconds while observing the battery terminal voltage. The voltage should remain above 10 volts the whole time the starter is engaged. After the key is released, the voltage should rise within about 5 seconds to about 95 percent of the original (pre-cranking) open circuit voltage. If the voltage fails to rise rapidly, the battery is defective and must be replaced. If the voltage drops below 10 volts at any time during the 15 seconds, recharge the battery and repeat the test. During the second test, if the voltage drops below 10 volts or if the voltage fails to recover rapidly, replace the battery.

To eliminate resistance in the starter circuit as the cause of the starting problem, or to isolate resistance in the starting circuit that is the cause of the starting problem, perform the following test. Connect the red lead of a voltmeter to the positive battery terminal and connect the black lead of that voltmeter to the biggest post of the starter solenoid, then crank the engine while observing the voltage. If the voltage displayed on the voltmeter is greater than 0.2 volts, then the cable has high resistance and could cause the current flow to be reduced enough to cause difficult starting. Repeat this procedure for each of the cables in the starter system. Replace any cable that shows a voltage drop greater than 0.2 volts.

This schematic shows the pin-out configuration of the OBD-2 diagnostic connector.

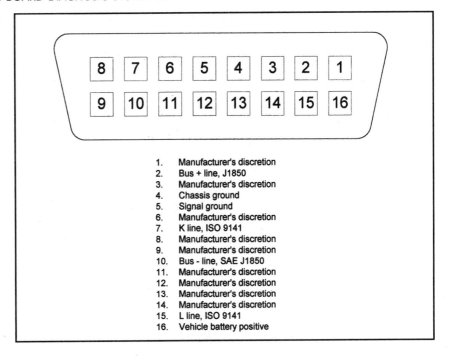

1. Manufacturer's discretion
2. Bus + line, J1850
3. Manufacturer's discretion
4. Chassis ground
5. Signal ground
6. Manufacturer's discretion
7. K line, ISO 9141
8. Manufacturer's discretion
9. Manufacturer's discretion
10. Bus - line, SAE J1850
11. Manufacturer's discretion
12. Manufacturer's discretion
13. Manufacturer's discretion
14. Manufacturer's discretion
15. L line, ISO 9141
16. Vehicle battery positive

Extended start

An extended start problem is where the engine cranks at the proper speed and starts normally, except that it must crank for several seconds first. The most common cause of this symptom is a loss of fuel volume in the system while the engine is shut off. The fuel pump is equipped with a check valve on the outlet side, which prevents fuel in the system from draining back into the tank. If the check valve should become defective, it will cause the extended start symptom. After the car sits for a while the engine will have to be cranked for several seconds (during which time it builds fuel pressure) before it will start. The cure for this problem is replacing the pump. Keep in mind that a loss of residual fuel volume is not the only possible cause of an extended start symptom, nor is the fuel pump the only cause of a loss of fuel volume or pressure.

A second and equally common cause of extended starting that is unique to GM fuel injection systems is the fuel pump relay. The fuel pump is activated by the fuel pump relay when the engine is being cranked. Once oil pressure is obtained, an oil pressure-controlled sending unit delivers voltage to the fuel pump. When the fuel pump relay goes bad, the engine will still start, but not until the engine has built up oil pressure.

Symptom Analysis Chart
Some of the Things to Check

Symptom	Possible "Electronic" Causes
No start	• Bad fuel pump/relay/fuse
	• Crankshaft sensor
	• Camshaft sensor
Start then die	• Oil pressure switch
	• Dirty throttle body
	• Defective idle controlstepper motor
Low idle cold	• Defective idle control stepper motor
	• Defective coolant sensor circuit
	• Dirty throttle bore
High idle cold	• Defective idle control stepper motor
	• Defective coolant sensor circuit
	• Vacuum leak
Rough idle	• Secondary ignition problems
	• Vacuum leak
	• Bad MAP/MAF sensor
Incorrect idle speed warm	• Defective idle control stepper motor
Hesitation problems	• Secondary ignition
	• TPS
	• Incorrect fuel pressure
	• Bad MAP/MAF sensor
	• Incorrect fuel pressure
Stall on deceleration	• Dirty throttle body
	• Defective idle control stepper motor
	• Bad MAP/MAF sensor
	• Bad coolant sensor
Runs-on (diesels)	• Leaking injector
	• Bad fuel pressure regulator
	• Bad canister purge valve

OBD-2 Code list: Powertrain Diagnostic Trouble Codes

P01XX Fuel and Air Metering

P0100	Mass or Volume Air Flow Circuit Malfunction
P0101	Mass or Volume Air Flow Circuit Range/Performance Problem
P0102	Mass or Volume Air Flow Circuit Low Input
P0103	Mass or Volume Air Flow Circuit High Input
P0104	Mass or Volume Air Flow Circuit Intermittent
P0105	Manifold Absolute Pressure/Barometric Pressure Circuit Malfunction
P0106	Manifold Absolute Pressure/Barometric Pressure Circuit Range/Performance Problem
P0107	Manifold Absolute Pressure/Barometric Pressure Circuit Low Input
P0108	Manifold Absolute Pressure/Barometric Pressure Circuit High Input
P0109	Manifold Absolute Pressure/Barometric Pressure Circuit Intermittent
P0110	Intake Air Temperature Circuit Malfunction
P0111	Intake Air Temperature Circuit Range/Performance Problem
P0112	Intake Air Temperature Circuit Low Input
P0113	Intake Air Temperature Circuit High Input
P0114	Intake Air Temperature Circuit Intermittent
P0115	Engine Coolant Temperature Circuit Malfunction
P0116	Engine Coolant Temperature Circuit Range/Performance Problem
P0117	Engine Coolant Temperature Circuit Low Input
P0118	Engine Coolant Temperature Circuit High Input
P0119	Engine Coolant Temperature Circuit Intermittent
P0120	Throttle/Pedal Position Sensor/Switch A Circuit Malfunction
P0121	Throttle/Pedal Position Sensor/Switch A Circuit Range/Performance Problem
P0122	Throttle/Pedal Position Sensor/Switch A Circuit Low Input
P0123	Throttle/Pedal Position Sensor/Switch A Circuit High Input
P0124	Throttle/Pedal Position Sensor/Switch A Circuit Intermittent
P0125	Insufficient Coolant Temperature for Closed Loop Control
P0126	Insufficient Coolant Temperature for Stable Operation
P0130	O2 Sensor Circuit Malfunction (Bank 1 Sensor 1)
P0131	O2 Sensor Circuit Low Voltage (Bank 1 Sensor 1)
P0132	O2 Sensor Circuit High Voltage (Bank 1 Sensor 1)
P0133	O2 Sensor Circuit Slow Response (Bank 1 Sensor 1)
P0134	O2 Sensor Circuit No Activity Detected (Bank 1 Sensor 1)
P0135	O2 Sensor Heater Circuit Malfunction
P0136	O2 Sensor Circuit Malfunction (Bank 1 Sensor 2)
P0137	O2 Sensor Circuit Low Voltage (Bank 1 Sensor 2)
P0138	O2 Sensor Circuit High Voltage (Bank 1 Sensor 2)
P0139	O2 Sensor Circuit Slow Response (Bank 1 Sensor 2)
P0140	O2 Sensor Circuit No Activity Detected (Bank 1 Sensor 2)
P0141	O2 Heater Circuit Malfunction (Bank 1 Sensor 2)
P0142	O2 Sensor Circuit Malfunction (Bank 1 Sensor 3)
P0143	O2 Sensor Circuit Low Voltage (Bank 1 Sensor 3)
P0144	O2 Sensor Circuit High Voltage (Bank 1 Sensor 3)
P0145	O2 Sensor Circuit Slow Response (Bank 1 Sensor 3)
P0146	O2 Sensor Circuit No Activity Detected (Bank 1 Sensor 3)
P0147	O2 Heater Circuit Malfunction (Bank 1 Sensor 3)
P0150	O2 Sensor Circuit Malfunction (Bank 2 Sensor 1)
P0151	O2 Sensor Circuit Low Voltage (Bank 2 Sensor 1)
P0152	O2 Sensor Circuit High Voltage (Bank 2 Sensor 1)
P0153	O2 Sensor Circuit Slow Response (Bank 2 Sensor 1)
P0154	O2 Sensor Circuit No Activity Detected (Bank 2 Sensor 1)
P0155	O2 Heater Circuit Malfunction (Bank 2 Sensor 1)
P0156	O2 Sensor Circuit Malfunction (Bank 2 Sensor 2)
P0157	O2 Sensor Circuit Low Voltage (Bank 2 Sensor 2)
P0158	O2 Sensor Circuit High Voltage (Bank 2 Sensor 2)
P0159	O2 Sensor Circuit Slow Response (Bank 2 Sensor 2)
P0160	O2 Sensor Circuit No Activity Detected (Bank 2 Sensor 2)
P0161	O2 Heater Circuit Malfunction (Bank 2 Sensor 2)
P0162	O2 Sensor Circuit Malfunction (Bank 2 Sensor 3)
P0163	O2 Sensor Circuit Low Voltage (Bank 2 Sensor 3)
P0164	O2 Sensor Circuit High Voltage (Bank 2 Sensor 3)
P0165	O2 Sensor Circuit Slow Response (Bank 2 Sensor 3)
P0166	O2 Sensor Circuit No Activity Detected (Bank 2 Sensor 3)
P0167	O2 Heater Circuit Malfunction (Bank 2 Sensor 3)
P0170	Fuel Trim Malfunction (Bank 1)
P0171	System too Lean (Bank 1)
P0172	System too Rich (Bank 1)
P0173	Fuel Trim Malfunction (Bank 2)
P0174	System too Lean (Bank 2)
P0175	System too Rich (Bank 2)
P0176	Fuel Composition Sensor Circuit Malfunction
P0177	Fuel Composition Sensor Circuit Range/Performance
P0178	Fuel Composition Sensor Circuit Low Input
P0179	Fuel Composition Sensor Circuit High Input
P0180	Fuel Temperature Sensor A Circuit Malfunction
P0181	Fuel Temperature Sensor A Circuit Range/Performance
P0182	Fuel Temperature Sensor A Circuit Low Input
P0183	Fuel Temperature Sensor A Circuit High Input
P0184	Fuel Temperature Sensor A Circuit Intermittent

P0185   Fuel Temperature Sensor B Circuit Malfunction
P0186   Fuel Temperature Sensor B
         Circuit Range/Performance
P0187   Fuel Temperature Sensor B Circuit Low Input
P0188   Fuel Temperature Sensor B Circuit High Input
P0189   Fuel Temperature Sensor B Circuit Intermittent
P0190   Fuel Rail Pressure Sensor Circuit Malfunction
P0191   Fuel Rail Pressure Sensor Circuit
         Range/Performance
P0192   Fuel Rail Pressure Sensor Circuit Low Input
P0193   Fuel Rail Pressure Sensor Circuit High Input
P0194   Fuel Rail Pressure Sensor Circuit Intermittent
P0195   Engine Oil Temperature Sensor Malfunction
P0196   Engine Oil Temperature Sensor
         Range/Performance
P0197   Engine Oil Temperature Sensor Low
P0198   Engine Oil Temperature Sensor High
P0199   Engine Oil Temperature Sensor Intermittent

P02XX Fuel and Air Metering

P0200   Injector Circuit Malfunction
P0201   Injector Circuit Malfunction - Cylinder 1
P0202   Injector Circuit Malfunction - Cylinder 2
P0203   Injector Circuit Malfunction - Cylinder 3
P0204   Injector Circuit Malfunction - Cylinder 4
P0205   Injector Circuit Malfunction - Cylinder 5
P0206   Injector Circuit Malfunction - Cylinder 6
P0207   Injector Circuit Malfunction - Cylinder 7
P0208   Injector Circuit Malfunction - Cylinder 8
P0209   Injector Circuit Malfunction - Cylinder 9
P0210   Injector Circuit Malfunction - Cylinder 10
P0211   Injector Circuit Malfunction - Cylinder 11
P0212   Injector Circuit Malfunction - Cylinder 12
P0213   Cold Start Injector 1 Malfunction
P0214   Cold Start Injector 2 Malfunction
P0215   Engine Shutoff Solenoid Malfunction
P0216   Injection Timing Control Circuit Malfunction
P0217   Engine Overtemp Condition
P0218   Transmission Over Temperature Condition
P0219   Engine Overspeed Condition
P0220   Throttle/Pedal Position Sensor/
         Switch B Circuit Malfunction
P0221   Throttle/Pedal Position Sensor/
         Switch B Circuit Range/Performance Problem
P0222   Throttle/Pedal Position Sensor/
         Switch B Circuit Low Input
P0223   Throttle/Pedal Position Sensor/
         Switch B Circuit High Input
P0224   Throttle/Pedal Position Sensor/
         Switch B Circuit Intermittent
P0225   Throttle/Pedal Position Sensor/
         Switch C Circuit Malfunction
P0226   Throttle/Pedal Position Sensor/
         Switch C Circuit Range/Performance Problem
P0227   Throttle/Pedal Position Sensor/
         Switch C Circuit Low Input

P0228   Throttle/Pedal Position Sensor/
         Switch C Circuit High Input
P0229   Throttle/Pedal Position Sensor/
         Switch C Circuit Intermittent
P0230   Fuel Pump Primary Circuit Malfunction
P0231   Fuel Pump Secondary Circuit Low
P0232   Fuel Pump Secondary Circuit High
P0233   Fuel Pump Secondary Circuit Intermittent
P0235   Turbocharger Boost Sensor A Circuit Malfunction
P0236   Turbocharger Boost Sensor A
         Circuit Range/Performance
P0237   Turbocharger Boost Sensor A Circuit Low
P0238   Turbocharger Boost Sensor A Circuit High
P0239   Turbocharger Boost Sensor B Malfunction
P0240   Turbocharger Boost Sensor B
         Circuit Range/Performance
P0241   Turbocharger Boost Sensor B Circuit Low
P0242   Turbocharger Boost Sensor B Circuit High
P0243   Turbocharger Wastegate Solenoid A Malfunction
P0244   Turbocharger Wastegate
         Solenoid A Range/Performance
P0245   Turbocharger Wastegate Solenoid A Low
P0246   Turbocharger Wastegate Solenoid A High
P0247   Turbocharger Wastegate Solenoid B Malfunction
P0248   Turbocharger Wastegate
         Solenoid B Range/Performance
P0249   Turbocharger Wastegate Solenoid B Low
P0250   Turbocharger Wastegate Solenoid B High
P0251   Injection Pump A Rotor/Cam Malfunction
P0252   Injection Pump A Rotor/Cam
         Range/Performance
P0253   Injection Pump A Rotor/Cam Low
P0254   Injection Pump A Rotor/Cam High
P0255   Injection Pump A Rotor/Cam Intermittent
P0256   Injection Pump B Rotor/Cam Malfunction
P0257   Injection Pump B Rotor/Cam Range/Performance
P0258   Injection Pump B Rotor/Cam Low
P0259   Injection Pump B Rotor/Cam High
P0260   Injection Pump B Rotor/Cam Intermittent
P0261   Cylinder 1 Injector Circuit Low
P0262   Cylinder 1 Injector Circuit High
P0263   Cylinder 1 Contribution/Balance Fault
P0264   Cylinder 2 Injector Circuit Low
P0265   Cylinder 2 Injector Circuit High
P0266   Cylinder 2 Contribution/Balance Fault
P0267   Cylinder 3 Injector Circuit Low
P0268   Cylinder 3 Injector Circuit High
P0269   Cylinder 3 Contribution/Balance Fault
P0270   Cylinder 4 Injector Circuit Low
P0271   Cylinder 4 Injector Circuit High
P0272   Cylinder 4 Contribution/Balance Fault
P0273   Cylinder 5 Injector Circuit Low
P0274   Cylinder 5 Injector Circuit High
P0275   Cylinder 5 Contribution/Balance Fault
P0276   Cylinder 6 Injector Circuit Low
P0277   Cylinder 6 Injector Circuit High

P0278	Cylinder 6 Contribution/Balance Fault		P0333	Knock Sensor 2 Circuit High Input (Bank 2)
P0279	Cylinder 7 Injector Circuit Low		P0334	Knock Sensor 2 Circuit Intermittent (Bank 2)
P0280	Cylinder 7 Injector Circuit High		P0335	Crankshaft Position Sensor A Circuit Malfunction
P0281	Cylinder 7 Contribution/Balance Fault		P0336	Crankshaft Position Sensor A Circuit Range/Performance
P0282	Cylinder 8 Injector Circuit Low		P0337	Crankshaft Position Sensor A Circuit Low Input
P0283	Cylinder 8 Injector Circuit High		P0338	Crankshaft Position Sensor A Circuit High Input
P0284	Cylinder 8 Contribution/Balance Fault		P0339	Crankshaft Position Sensor A Circuit Intermittent
P0285	Cylinder 9 Injector Circuit Low		P0340	Camshaft Position Sensor Circuit Malfunction
P0286	Cylinder 9 Injector Circuit High		P0341	Camshaft Position Sensor Circuit Range/Performance
P0287	Cylinder 9 Contribution/Balance Fault		P0342	Camshaft Position Sensor Circuit Low Input
P0288	Cylinder 10 Injector Circuit Low		P0343	Camshaft Position Sensor Circuit High Input
P0289	Cylinder 10 Injector Circuit High		P0344	Camshaft Position Sensor Circuit Intermittent
P0290	Cylinder 10 Contribution/Balance Fault		P0350	Ignition Coil Primary/Secondary Circuit Malfunction
P0291	Cylinder 11 Injector Circuit Low		P0351	Ignition Coil A Primary/ Secondary Circuit Malfunction
P0292	Cylinder 11 Injector Circuit High			
P0293	Cylinder 11 Contribution/Balance Fault		P0352	Ignition Coil B Primary/ Secondary Circuit Malfunction
P0294	Cylinder 12 Injector Circuit Low			
P0295	Cylinder 12 Injector Circuit High		P0353	Ignition Coil C Primary/ Secondary Circuit Malfunction
P0296	Cylinder 12 Contribution/Range Fault			

P03XX Ignition System or Misfire

P0300	Random/Multiple Cylinder Misfire Detected		P0354	Ignition Coil D Primary/ Secondary Circuit Malfunction
P0301	Cylinder 1 Misfire Detected		P0355	Ignition Coil E Primary/ Secondary Circuit Malfunction
P0302	Cylinder 2 Misfire Detected			
P0303	Cylinder 3 Misfire Detected		P0356	Ignition Coil F Primary/ Secondary Circuit Malfunction
P0304	Cylinder 4 Misfire Detected			
P0305	Cylinder 5 Misfire Detected		P0357	Ignition Coil G Primary/ Secondary Circuit Malfunction
P0306	Cylinder 6 Misfire Detected			
P0307	Cylinder 7 Misfire Detected		P0358	Ignition Coil H Primary/ Secondary Circuit Malfunction
P0308	Cylinder 8 Misfire Detected			
P0309	Cylinder 9 Misfire Detected		P0359	Ignition Coil I Primary/ Secondary Circuit Malfunction
P0310	Cylinder 10 Misfire Detected			
P0311	Cylinder 11 Misfire Detected		P0360	Ignition Coil J Primary/ Secondary Circuit Malfunction
P0312	Cylinder 12 Misfire Detected			
P0320	Ignition/Distributor Engine Speed Input Circuit Malfunction		P0361	Ignition Coil K Primary/ Secondary Circuit Malfunction
P0321	Ignition/Distributor Engine Speed Input Circuit Range/Performance		P0362	Ignition Coil L Primary/ Secondary Circuit Malfunction
P0322	Ignition/Distributor Engine Speed Input Circuit No Signal		P0370	Timing Reference High Resolution Signal A Malfunction
P0323	Ignition/Distributor Engine Speed Input Circuit Intermittent		P0371	Timing Reference High Resolution Signal A Too Many Pulses
P0325	Knock Sensor 1 Circuit Malfunction (Bank 1 or Single Sensor)		P0372	Timing Reference High Resolution Signal A Too Few Pulses
P0326	Knock Sensor 1 Circuit Range/Performance (Bank 1 or Single Sensor)		P0373	Timing Reference High Resolution Signal A Intermittent/Erratic Pulses
P0327	Knock Sensor 1 Circuit Low Input (Bank 1 or Single Sensor)		P0374	Timing Reference High Resolution Signal A No Pulses
P0328	Knock Sensor 1 Circuit High Input (Bank 1 or Single Sensor)		P0375	Timing Reference High Resolution Signal B Malfunction
P0329	Knock Sensor 1 Circuit Intermittent (Bank 1 or Single Sensor)		P0376	Timing Reference High Resolution Signal B Too Many Pulses
P0330	Knock Sensor 2 Circuit Malfunction (Bank 2)			
P0331	Knock Sensor 2 Circuit Range/Performance (Bank 2)			
P0332	Knock Sensor 2 Circuit Low Input (Bank 2)			

P0377 Timing Reference High
    Resolution Signal B Too Few Pulses

P0378 Timing Reference High Resolution Signal B
    Intermittent/Erratic Pulses

P0379 Timing Reference High Resolution Signal B
    No Pulses

P0380 Glow Plug/Heater Circuit Malfunction

P0381 Glow Plug/Heater Indicator Circuit Malfunction

P0385 Crankshaft Position Sensor B
    Circuit Malfunction

P0386 Crankshaft Position Sensor B
    Circuit Range/Performance

P0387 Crankshaft Position Sensor B Circuit Low Input

P0388 Crankshaft Position Sensor B Circuit High Input

P0389 Crankshaft Position Sensor B Circuit Intermittent

P04XX Auxiliary Emission Controls

P0400 Exhaust Gas Recirculation Flow Malfunction

P0401 Exhaust Gas Recirculation Flow
    Insufficient Detected

P0402 Exhaust Gas Recirculation Flow
    Excessive Detected

P0403 Exhaust Gas Recirculation Circuit
    Malfunction

P0404 Exhaust Gas Recirculation Circuit
    Range/Performance

P0405 Exhaust Gas Recirculation Sensor A Circuit Low

P0406 Exhaust Gas Recirculation Sensor A Circuit High

P0407 Exhaust Gas Recirculation Sensor B Circuit Low

P0408 Exhaust Gas Recirculation Sensor B Circuit High

P0410 Secondary Air Injection System Malfunction

P0411 Secondary Air Injection System
    Incorrect Flow Detected

P0412 Secondary Air Injection System
    Switching Valve A Circuit Malfunction

P0413 Secondary Air Injection System
    Switching Valve A Circuit Open

P0414 Secondary Air Injection System
    Switching Valve A Circuit Shorted

P0415 Secondary Air Injection System
    Switching Valve B Circuit Malfunction

P0416 Secondary Air Injection System
    Switching Valve B Circuit Open

P0417 Secondary Air Injection System
    Switching Valve B Circuit Shorted

P0420 Catalyst System Efficiency
    Below Threshold (Bank 1)

P0421 Warm Up Catalyst Efficiency
    Below Threshold (Bank 1)

P0422 Main Catalyst Efficiency Below
    Threshold (Bank 1)

P0423 Heated Catalyst Efficiency
    Below Threshold (Bank 1)

P0424 Heated Catalyst Temperature
    Below Threshold (Bank 1)

P0430 Catalyst System Efficiency
    Below Threshold (Bank 2)

P0431 Warm Up Catalyst Efficiency
    Below Threshold (Bank 2)

P0432 Main Catalyst Efficiency Below
    Threshold (Bank 2)

P0433 Heated Catalyst Efficiency
    Below Threshold (Bank 2)

P0434 Heated Catalyst Temperature
    Below Threshold (Bank 2)

P0440 Evaporative Emission Control System Malfunction

P0441 Evaporative Emission Control
    System Incorrect Purge Flow

P0442 Evaporative Emission Control
    System Leak Detected (small leak)

P0443 Evaporative Emission Control
    System Purge Control Valve Circuit Malfunction

P0444 Evaporative Emission Control
    System Purge Control Valve Circuit Open

P0445 Evaporative Emission Control
    System Purge Control Valve Circuit Shorted

P0450 Evaporative Emission Control
    System Pressure Sensor Malfunction

P0451 Evaporative Emission Control
    System Pressure Sensor Range/Performance

P0452 Evaporative Emission Control
    System Pressure Sensor Low Input

P0453 Evaporative Emission Control
    System Pressure Sensor High Input

P0454 Evaporative Emission Control
    System Pressure Sensor Intermittent

P0455 Evaporative Emission Control
    System Leak Detected (gross leak)

P0460 Fuel Level Sensor Circuit Malfunction

P0461 Fuel Level Sensor Circuit Range/Performance

P0462 Fuel Level Sensor Circuit Low Input

P0463 Fuel Level Sensor Circuit High Input

P0464 Fuel Level Sensor Circuit Intermittent

P0465 Purge Flow Sensor Circuit Malfunction

P0466 Purge Flow Sensor Circuit Range/Performance

P0467 Purge Flow Sensor Circuit Low Input

P0468 Purge Flow Sensor Circuit High Input

P0469 Purge Flow Sensor Circuit Intermittent

P0470 Exhaust Pressure Sensor Malfunction

P0471 Exhaust Pressure Sensor Range/Performance

P0472 Exhaust Pressure Sensor Low Input

P0473 Exhaust Pressure Sensor High Input

P0474 Exhaust Pressure Sensor Intermittent

P0475 Exhaust Pressure Control Valve Malfunction

P0476 Exhaust Pressure Control Valve
    Range/Performance

P0477 Exhaust Pressure Control Valve Low Input

P0478 Exhaust Pressure Control Valve High Input

P0479 Exhaust Pressure Control Valve Intermittent

P05XX Vehicle Speed, Idle Control, and Auxiliary Inputs

P0500	Vehicle Speed Sensor Malfunction
P0501	Vehicle Speed Sensor Range/Performance
P0502	Vehicle Speed Sensor Low Input
P0503	Vehicle Speed Sensor Intermittent/Erratic/High
P0505	Idle Control System Malfunction
P0506	Idle Control System RPM Lower Than Expected
P0507	Idle Control System RPM Higher Than Expected
P0510	Closed Throttle Position Switch Malfunction
P0530	A/C Refrigerant Pressure Sensor Circuit Malfunction
P0531	A/C Refrigerant Pressure Sensor Circuit Range/Performance
P0532	A/C Refrigerant Pressure Sensor Circuit Low Input
P0533	A/C Refrigerant Pressure Sensor Circuit High Input
P0534	Air Conditioner Refrigerant Charge Loss
P0550	Power Steering Pressure Sensor Circuit Malfunction
P0551	Power Steering Pressure Sensor Circuit Range/Performance
P0552	Power Steering Pressure Sensor Circuit Low Input
P0553	Power Steering Pressure Sensor Circuit High Input
P0554	Power Steering Pressure Sensor Circuit Intermittent
P0560	System Voltage Malfunction
P0561	System Voltage Unstable
P0562	System Voltage Low
P0563	System Voltage High
P0565	Cruise Control On Signal Malfunction
P0566	Cruise Control Off Signal Malfunction
P0567	Cruise Control Resume Signal Malfunction
P0568	Cruise Control Set Signal Malfunction
P0569	Cruise Control Coast Signal Malfunction
P0570	Cruise Control Accel Signal Malfunction
P0571	Cruise Control/Brake Switch A Circuit Malfunction
P0572	Cruise Control/Brake Switch A Circuit Low
P0573	Cruise Control/Brake Switch A Circuit High
P0574	through P0580 Reserved for Cruise Codes

P06XX Computer and Auxiliary Outputs

P0600	Serial Communication Link Malfunction
P0601	Internal Control Module Memory Check Sum Error
P0602	Control Module Programming Error
P0603	Internal Control Module Keep Alive Memory (KAM) Error
P0604	Internal Control Module Random Access Memory (RAM) Error
P0605	Internal Control Module Read Only Memory (ROM) Error
P0606	PCM Processor Fault

OBD-2 Data Stream

The data stream for OBD-2 is very similar to OBD-1 for GM applications

Fuel and Air Metering
Mass Air Flow

The Mass Air Flow sensor (MAF) measures the volume and density of the air entering the engine. The sensor uses a heated element and a thermistor to measure the air mass. As the air passes through the sensor it comes in contact with the thermistor. A module mounted on the side of the sensor then heats the heated element to about 100 degrees above the temperature of incoming air. As the air passes into the air induction system, it cools the heated element; the amount of cooling is directly proportional to the volume of air entering the engine and inversely proportional to the temperature of the air entering the engine. This information is then processed by the computer and used to calculate the proper injector on-time and ignition timing. Additionally, the reading from the MAF is sent down the serial data stream to the scan tool.

Typically, the scanner will be programmed to read the MAF signal in grams per second. A good reading in this data field on a 3.8-liter engine is about 10 grams per second at an idle, a little more on a bigger engine and a little less on a smaller engine. Some scanners will also produce a corollary data field that indicates the volumetric flow rate in liters per minute or milliliters per second.

Manifold Absolute Pressure

The MAP sensor measures the pressure in the intake manifold. Technicians usually think of the intake manifold as having a vacuum in it. In truth, the manifold just has a pressure that is less than atmospheric pressure, so it appears to have a vacuum. When the engine is not running, there is full atmospheric pressure in the manifold—about 30 inches of mercury, which is about 15 pounds per square inch, or about 104 kiloPascals at sea level. When the engine is idling, the pressure in the manifold is approximately one-third of the atmospheric pressure.

The PCM uses the signal from the MAP sensor to determine the load on the engine. As engine load increases, the manifold vacuum tends to drop, which means that the pressure is actually rising. In general, the higher the pressure, the greater the load on the engine. On applications that do not use a Mass Air Flow sensor, the MAP sensor is also used to measure air flow. This method of measuring air flow is called the Speed/Density method and involves comparing the pressure in the manifold to atmospheric pressure. The closer the manifold pressure is to atmospheric pressure, the greater the air flow into the engine is presumed to be. The greater the difference in these pressures, the less air the PCM presumes to be flowing into the engine.

MAP sensor readings on many scanners will be in kiloPascals, therefore, at sea-level, when the engine is not running, the reading will be about 104 kiloPascals or about 14.7 psi, which is about 29–30 inches of mercury. A higher altitude, or a day with lower atmospheric pressure, will yield a lower reading. If you are visiting Death Valley and decide to check your MAP sensor readings, expect the key-on-engine-off readings to be lower than at sea level. When the engine starts and idles, the pressure in the manifold drops to about one-third of its key-on/engine-off pressure, so, likewise, the kiloPascals reading will drop to about one-third of its original key-on/engine-off reading.

Intake Air Temperature

The Intake Air Temperature sensor is used by the PCM to determine the density of the air entering the engine, which is essential when trying to measure the mass of the air entering the engine. When a Mass Air Flow sensor is used, it alone may be integrated into the system to measure the density of the air, without the assistance of an air temperature sensor.

The units of measure for the air temperature sensor is usually degrees Celsius. Some scanners will translate this reading to degrees Fahrenheit. When considering the appropriateness of the reading, consider the ambient temperature, the underhood temperature, and the location of the sensor. Generally speaking, if the air temperature sensor is located in the air cleaner, the reading should always be within a few degrees of ambient. If the sensor is located in the intake manifold, the temperature will be within a few degrees of ambient at times of high air flow, but will tend to rise as the air flow increases.

Engine Coolant Temperature

The engine coolant temperature sensor provides information to the computer for the control air/fuel ratio, ignition timing and emission control devices. On the OBD-2 scanner the reading will be in either degrees Celsius or degrees Fahrenheit. Naturally, the temperature indicated should be close to the assumed temperature of the engine.

Throttle Position Volts

The Throttle Position Sensor (TPS) informs the computer of changes in power requested by the driver. When the computer sees the TPS voltage increasing, it increases the injector on-time to enrich the mixture for better acceleration.

When the throttle is closed, a typical reading would be in the neighborhood of 0.5 volts. As the throttle is opened, the voltage should increase proportionally, peaking at between 4.0 and 4.5 volts when the throttle is wide open.

Throttle Position Percent

Most OBD-2 scanners will also present the throttle position sensor readings in percent form. When the throttle is closed, the reading should be 0 percent. As the throttle is opened, the reading should rise proportionally to 100 percent.

Loop Status

This data field reports whether or not the computer is accepting and using the information from the oxygen sensor. When the data field indicates that the system is operating in Open Loop, the computer is ignoring the oxygen sensor. When the system is operating in Closed Loop mode, the computer is accepting and using the information from the oxygen sensor to fine tune the air/fuel ratio.

O2 - Bank 1, Sensor 1

The oxygen sensor monitors the exhaust oxygen content of an OBD-2 application just like in earlier fuel-injected engines. The oxygen sensor voltage is displayed in a data field on the OBD-2 scanner. The voltage range of the oxygen sensor is from 100 to 900 millivolts. When the voltage from the sensor is greater than 450 millivolts, it means that the oxygen content of the exhaust is low, therefore the engine is believed to be running rich. The voltage from the oxygen sensor should be constantly toggling above and below 450 millivolts.

O2 - Bank 1, Sensor 2

OBD-2 has been designed to recognize the fact that there is a need to monitor the efficiency of the catalytic converter. Therefore a second oxygen sensor will be found on most OBD-2 application. This oxygen sensor should also be fluctuating.

O2 - Bank 1, Sensor 3

Should there be a second catalytic converter, it would be necessary to monitor its efficiency as well. Therefore a *third* oxygen sensor may be located downstream of the second catalytic converter.

O2 - Bank 2, Sensor 1, O2 - Bank 2, Sensor 2, O2 - Bank 2, Sensor 3

Some "V" (V-6, V-8, V-10, etc.) engines may utilize a second group of oxygen sensors on the second engine bank. For this reason there may be as many as *six* oxygen sensors on a single engine.

Short Term Fuel Trim, Bank 1

In the OBD-1 Chevrolet applications, the ECM had the ability to adjust for errors in air/fuel ratio by altering its "integrator." The integrator was a program of the ECM that responded to changes in the oxygen content of the exhaust. When the engine was warm and the computer was operating in closed loop, the oxygen sensor would report changes in air/fuel ratio. If the oxygen sensor voltage was low, the computer would respond by increasing the Integrator number from the norm of 128 to a higher number. If the oxygen sensor voltage was high, the computer responded by decreasing the Integrator number. The normal range was 118 to 138. Therefore a number greater than 138 indicated that the ECM was adding fuel to compensate for a perceived lean operating condition. A number less than 118 meant that the computer was decreasing the delivery of fuel to compensate for a rich running condition. The theoretical, though not necessarily actual, range was 0–255.

With OBD-2 the name of this process and the units of measurement change, but the process itself is identical. When the PCM is not responding to perceived errors in the air/fuel ratio indicated by the exhaust oxygen content, the reading in the Short Term Fuel Trim data field will be 0 percent (equivelant of a 128 Integrator). If the PCM is richening the mixture in response to lean exhaust conditions, the data field will display a positive percentage reading. If the PCM is leaning the mixture, the data field will display negative percentage readings.

Short Term Fuel Trim, Bank 2

Those applications that have oxygen sensors on each side of the engine will have separate Short Term Fuel Trim programs for each side.

Long Term Fuel Trim, Bank 1

The Short Term Fuel Trim only corrects for perceived air fuel ratio problems when the engine is warm and operating in closed loop. In the old OBD-1 system, the Block Learn program would store information about air/fuel ratio problems and correct for these problem based on memory, even before the oxygen sensor and engine are warm enough to enter closed loop. The units of measure were identical to those of the Integrator.

For OBD-2, this function is handled by Long Term Fuel Trim. Like the Short Term Fuel Trim, the units of measure are positive and negative percentages.

Long Term Fuel Trim, Bank 2

Just like the Sort Term Fuel Trim, there is also a Long Term Fuel Trim for bank two of the engine.

Injector Pulse Width, Cylinders 1–12

At idle, when the engine is warmed up, most engines have an injector base pulse width of about 1–4 milliseconds. An important thing to keep in mind about this data field, is that it may not be indicating the actual on-time. The pulse width indicated in this field is selected based on the air flow and engine speed. Other sensors may cause the injector on-time to be longer than the base pulse width. Also, if the PCM is erroneously selecting the wrong injector on-time, it will not report that mistake to this data field.

Ignition System or Misfire

Cylinder Misfire, Cylinders 1–12

These data fields indicate whether or not the PCM is detecting fluctuations in crankshaft speed or crankshaft acceleration on each of up to 12 cylinders. These fluctuations are interpreted by the PCM as misfiring cylinders.

Cam signal

Typically, this data field will only read either "No" or "Yes." If the field says "yes," then it indicates that

the PCM has received a pulse from the camshaft sensor. The camshaft sensor signal is necessary to begin the sequencing of the fuel injectors. On some applications, if the computer does not receive this signal, then the injectors will not begin sequencing. On other applications, the engine will start but it will be running in a "limp-in" mode and will be firing all injectors simultaneously.

Knock Sensor 1, Knock Sensor 2

Knocking, or detonation, occurs when the temperatures in the combustion chamber remains high enough to ignite the air fuel charge fed into that combustion chamber on the next engine cycle before the spark plug fires. There are two strategies for reducing this temperature. The first, and most common, is to retard the ignition timing. The second is to slightly richen the mixture. The knock sensor is responsible for detecting these detonations so that the PCM can respond according to its programming. These data fields usually reflect only whether or not detonation is presently occurring.

Crankshaft Signal

This is an RPM signal.

Auxiliary Emission Controls
EGR Percentage

Since the early 1970s manufacturers have been controlling oxides of nitrogen emissions by metering a small amount of exhaust gases back into the combustion chamber. These exhaust gases act as a heat sink to cool the combustion process. In the early days, this process was difficult to control accurately. Often, the flow rate of the gases was excessive, which would cause the engine to stumble and hesitate. On late model applications, the EGR is controlled by the PCM. In some cases, the PCM controls a vacuum solenoid that, in turn, controls the EGR position. In many later models, the EGR itself is a stepper motor. This data field monitors that motor and informs the technician of the amount of opening of the

EGR. In most cases, this reading will be 0 percent at an idle; the percentage will increase when the vehicle is driven. In a few cases, the reading will be 100 percent at an idle and decrease when the vehicle is driven.

A/C High Side Pressure

This data field reports the pressure of the refrigerant on the high side of the air conditioning system.

Secondary Air Injection

Secondary air injection is known more commonly as the air pump system. This data field will report if the air is being pumped upstream of the oxygen sensor, downstream of the oxygen sensor, or into the air cleaner.

Catalyst Efficiency, Bank 1, Bank 2

The PCM monitors the exhaust oxygen content ahead of the converter and downstream of the converter. A comparison of the two readings allows the PCM to determine the efficiency of the catalytic converter. Refer to the operating manual for the scanner you are using to determine the units of measure being used.

Evaporative Canister Percentage

Generally speaking, the evaporative canister will be closed when the engine is idling. When the engine speed is above idle and the engine is under a light load (indicating cruise conditions), the PCM will open the canister purge valve to allow fuel vapor to flow into the intake. Once in the intake, the fuel vapor flows to the combustion chamber and are consumed. For most applications, the percentage should be very low or zero when the engine is idling and increase when the vehicle is driven.

Vehicle Speed Control and Idle Control

Vehicle Speed

This is the current vehicle speed, displayed in kilometers per hour (kph) or mile per hour (mph).

Idle Speed Control

The idle speed control device on most applications is a stepper motor. On OBD-1 applications, this data field reads in counts. This would be a reading between 0 and 255. Typically, the reading would be in the vicinity of 20 counts at an idle. Under OBS-2, the units of measure change to percentages; when the engine is idling, the reading on the scanner should be about 10–20 percent.

Transmission

Transmission Range Input (PRNDL)

This data field indicates the position of the gear selector lever.

Transmission Oil Temperature

This is the temperature of the transmission fluid. Depending on the scanner, the reading will be in either degrees Fahrenheit or degrees Celsius.

Transmission Input Speed

This is the speed, in RPM, of the input shaft speed of the transmission. Basically, this is engine RPM less the slippage of the torque converter. Monitoring this specification allows the PCM to diagnose the torque converter.

Transmission Output Speed

This field is also measured in RPM and is used by the PCM to diagnose slipping clutch packs in the transmission. The PCM compares transmission input speed to transmission output speed to make this determination.

Gear Ratio (Correct/Incorrect), Gears 1–5

This reports the PCM's interpretation of the relationship between the transmission input speed, the transmission output speed and the gear selector indication. The PCM has been programmed to believe that there is a very specific relationship between these items. When these relationships are incorrect, these data fields will report errors.

Torque Converter Clutch Lock-up

This data field reports the state of the lock-up torque converter.

Pressure Control Solenoid Percentage

Many of the electronically controlled automatic transmissions use a solenoid to control the operational pressure of the transmission. This reading should be in the mid range.

Pressure Control PSI

This data field reports the operational pressure of the automatic transmission to the technician. Refer to the specifications for the transmission you have to determine if the readings your get are correct.

Shift Solenoid Status, Solenoids A-E

Electronic automatic transmissions use solenoids to apply clutches and bands. These data fields report the current status of each of these solenoids. Refer to a troubleshooting guide for the transmission in question, to find out what the states should be in any given gear.

APPENDIX

Selected OBD-2 Abbreviations

Acronym/Recommended Terms	Abbr.	Definitions
Accelerator Pedal	AP	See Glossary Entry ACCELERATOR PEDAL.
Air Cleaner	ACL	See Glossary Entry CLEANER.
Air Conditioning	A/C	See Glossary Entry AIR CONDITIONING.
Automatic Transaxle	A/T	See Glossary Entry TRANSAXLE.
Automatic Transmission	A/T	See Glossary Entry TRANSMISSION.
Barometric Pressure	BARO	See Glossary Entry PRESSURE.
Battery Positive Voltage	B+	See Glossary Entry BATTERY.
Camshaft Position	CMP	See Glossary Entry CAMSHAFT.
Canister		See Glossary Entry CANISTER.
Carburetor	CARB	See Glossary Entry CARBURETOR.
Charge Air Cooler	CAC	A device which lowers the temperature of the pressure intake air.
Closed Loop	CL	See Glossary Entry CLOSED LOOP.
Closed Throttle Position	CTP	See Glossary Entry THROTTLE.
Clutch Pedal Position	CPP	See Glossary Entry CLUTCH.
Continuous Fuel Injection	CFI	A fuel injection system with the injector flow controlled by fuel pressure.
Continuous Trap Oxidizer	CTOX	A system for lowering diesel engine particulate emissions by collecting exhaust particulates and continuously burning them through oxidation.
Crankshaft Position	CKP	See Glossary Entry CRANKSHAFT.
Data Link Connector	DLC	Connector providing access and/for control of the vehicle information, operating conditions, and diagnostic information
Diagnostic Test Mode	DTM	A level of diagnostic capability in an On Board Diagnostic (OBD) system. This may include different functional states to observe signals, a base level to read diagnostic trouble codes, a monitor level which includes information on signal levels, bi-directional control with on/off board aids, and the ability to interface with remote diagnosis
Diagnostic Trouble Code	DTC	An alphanumeric identifier for a fault condition identified by the On Board Diagnostic System.
Direct Fuel Injection	DFI	Fuel injection system that supplies fuel directly into the combustion chamber.
Distributor Ignition	DI	A system in which the ignition coil secondary circuit is switched by a distributor in proper sequence to various spark plugs.

Early Fuel Evaporation	EFE	Enhancing air/fuel vaporization during engine warm up.
EGR Temperature	EGRT	Sensing EGR function based on temperature change. Primarily in systems with mechanical flow control devices.
Electrically Erasable Programmable	EEPROM	An electronic Read Only Memory device named electrically erasable programmable read only memory.
Electronic Ignition	EI	A system in which the ignition coil secondary circuit is dedicated to specific spark plugs without the use of a distributor.
Engine Control	EC	See Glossary Entries ENGINE and CONTROL.
Engine Control Module	ECM	See Glossary Entries ENGINE, CONTROL, MODULE.
Engine Coolant Level	ECL	See Glossary Entries ENGINE, COOLANT, LEVEL.
Engine Coolant Temperature	ECT	See Glossary Entries ENGINE, COOLANT.
Engine Modification	EM	A method of lowering engine emissions through changes in basic engine construction or in fuel and spark calibration.
Engine Speed	RPM	See Glossary Entries ENGINE, SPEED.
Erasable Programmable	EPROM	An electronic device named erasable programmable read only memory.
Evaporative Emission	EVAP	A system used to prevent fuel vapor from escaping into the atmosphere. Typically includes a charcoal canister to store fuel vapors.
Exhaust Gas Recirculation	EGR	Reducing NOx emissions levels by adding exhaust gas to the incoming air/fuel mixture.
Fan Control	FC	See Glossary Entries FAN, CONTROL.
Flash Electrically Erasable	FEEPROM	An electronic device named flash electrically erasable programmable read only memory.
Flash Erasable Programmable	FEPROM	An electronic device named flash erasable programmable read only memory.
Flexible Fuel	FF	A system capable of using a variety of fuels for vehicle operation.
Fourth Gear	4GR	Identifies the gear in which the transmission is operating in at a particular moment (e.g., the Transmission Range (TR) switch may indicate that DRIVE was selected, but the transmission is operating in fourth gear as indicated by 4GR switch).
Fuel Level Sensor		See Glossary Entries FUEL, SENSOR.
Fuel Pressure		See Glossary Entries FUEL, PRESSURE.
Fuel Pump	FP	See Glossary Entries FUEL, PUMP.
Fuel Trim	FT	A fuel correction term.
Generator	GEN	See Glossary Entry GENERATOR.
Governor		See Glossary Entry GOVERNOR.
Governor Control Module	GCM	See Glossary Entries GOVERNOR, CONTROL and MODULE.
Ground	GND	See Glossary Entry GROUND.
Heated Oxygen Sensor	HO2S	An oxygen sensor (O2S) that is electrically heated.
Idle Air Control	IAC	Electrical or mechanical control of throttle bypass air.
Idle Speed Control	ISC	Electronic control of minimum throttle position.
Ignition Control	IC	See Glossary Entries IGNITION and CONTROL.
Ignition Control Module	ICM	See Glossary Entries IGNITION, CONTROL, MODULE.
Indirect Fuel Injection	IFI	An injection system that supplies fuel into a combustion pre-chamber (Diesel).
Inertia Fuel Shutoff	IFS	An inertia system that shuts off the fuel delivery system when activated by predetermined force limits.
Intake Air	IA	See Glossary Entry INTAKE AIR.
Intake Air Temperature	IAT	See Glossary Entry INTAKE AIR.
Knock Sensor	KS	See Glossary Entries KNOCK, SENSOR.
Malfunction Indicator Lamp	MIL	A required on-board indicator to alert the driver of an emission-related malfunction.
Manifold Absolute Pressure	MAP	See Glossary Entries MANIFOLD, PRESSURE.
Manifold Differential Pressure	MDP	See Glossary Entries MANIFOLD, PRESSURE.

Manifold Surface Temperature	MST	See Glossary Entry MANIFOLD.
Manifold Vacuum Zone	MVZ	See Glossary Entries MANIFOLD, VACUUM.
Mass Air Flow	MAF	A system which provides information on the mass flow rate of the intake air to the engine.
Mixture Control	MC	A device which regulates bleed air, fuel, or both, on carbureted vehicles.
Multiport Fuel Injection	MFI	A fuel delivery system in which each cylinder is individually fueled.
Nonvolatile Random Access	NVRAM	An electronic device named nonvolatile Memory random access memory.
Oil Pressure		Pressure in the lubrication system.
On Board Diagnostic	OBD	A system that monitors some or all computer input and control signals. Signal(s) outside of the predetermined limits imply a fault in the system or in a related system.
Open Loop	OL	See Glossary Entry OPEN LOOP.
Oxidation Catalytic Converter	OC	A catalytic converter system that reduces levels of HC and CO.
Oxygen Sensor	O2S	A sensor that detects oxygen (O2) content in the exhaust gases.
Park/Neutral Position	PNP	See Glossary Entry PARK/NEUTRAL.
Periodic Trap Oxidizer	PTOX	A system for lowering diesel engine particulate emissions by collecting exhaust particulates and periodically burning them through oxidation.
Positive Crankcase Ventilation	PCV	Positive ventilation of crankcase emissions.
Power Steering Pressure	PSP	See Glossary Entry POWER STEERING.
Powertrain Control Module	PCM	See Glossary Entries POWERTRAIN, CONTROL, MODULE.
Programmable Read Only Memory	PROM	An electronic device named programmable (by the manufacturer) read only memory.
Pulsed Secondary Air Injection	PAIR	A pulse driven system for providing secondary air without an air pump by using the engine exhaust system pressure fluctuations or pulses.
Random Access Memory	RAM	An electronic device named random access memory.
Read Only Memory	ROM	An electronic device named read only memory.
Relay Module	RM	See Glossary Entries RELAY, MODULE.
Scan Tool	ST	See Glossary Entry SCAN TOOL.
Secondary Air Injection	AIR	A pump driven system for providing secondary air.
Sequential Multiport Fuel Injection	SFI	A multiport fuel delivery system in which each injector is individually energized and timed relative to its cylinder intake event. Normally fuel is delivered to each cylinder once per two crankshaft revolutions in four cycle engines and once per crankshaft revolution in two cycle engines.

Service Reminder Indicator	SRI	An indicator used to identify a service requirement.
Smoke Puff Limiter	SPL	A system to reduce diesel exhaust smoke during vehicle acceleration or gear changes.
Supercharger	SC	See Glossary Entry SUPERCHARGER.
Supercharger Bypass	SCB	See Glossary Entry SUPERCHARGER.
System Readiness Test	SRT	System readiness test as applicable to OBDII scan tool communications.
Thermal Vacuum Valve	TVV	A valve that controls vacuum levels or routing based on temperature.
Third Gear	3GR	Identifies the gear in which the transmission is operating in at a particular moment (e.g., the Transmission Range (TR) switch may indicate that drive was selected, but the transmission is operating in third gear as indicated by 3GR switch).
Three Way + Oxidation	TWC+OC	A catalytic converter system that has both Catalytic Converter Three Way Catalyst (TWC) and Oxidation Catalyst (OC). Usually secondary air is introduced between the two catalysts.
Three Way Catalytic Converter	TWC	A catalytic converter system that reduces levels of HC, CO, and NOx.
Throttle Body	TB	See Glossary Entries THROTTLE, BODY.
Throttle Body Fuel Injection	TBI	An electronically controlled fuel injection system in which one or more fuel injectors are located in a throttle body.
Throttle Position	TP	See Glossary Entry THROTTLE.
Torque Converter Clutch	TCC	See Glossary Entry CONVERTER.
Transmission Control Module	TCM	See Glossary Entries TRANSMISSION, CONTROL, MODULE.
Transmission Range	TR	See Glossary Entries TRANSMISSION, RANGE.
Turbocharger	TC	See Glossary Entry TURBOCHARGER.
Vehicle Speed Sensor	VSS	A sensor which provides vehicle speed information.
Voltage Regulator	VR	See Glossary Entry REGULATOR.
Volume Air Flow	VAF	A system which provides information on the volume flow rate of the intake air to the engine.
Warm Up Oxidation	WU-OC	A catalytic converter system designed to Catalytic Converter lower HC and CO emissions during engine warm up. Usually located in or near the exhaust manifold.
Warm Up Three Way	WU-TWC	A catalytic converter system designed to Catalytic Converter lower HC, CO, and NOx emissions during engine warm up. Usually located in or near the exhaust manifold.
Wide Open Throttle	WOT	See Glossary Entry THROTTLE.

Minimum Air and TPS Specifications

Year	Model	Engine (liters)	VIN #	Injection system	TPS (volts)	Minimum Air (rpm)
1982	Citation	2.5	2	TBI	<1.25	500±25
	Celebrity	2.5	2	TBI	<1.25	500±25
	Camaro	2.5	2	TBI	<1.25	500±25
	Camaro	5.0	S	CFI	0.525±0.075	400
	Corvette	5.7	8	CFI	0.525±0.075	400
1983	Cavalier	2.0	P	TBI	<1.25	650±25
	Cavalier	2.5	2	TBI	<1.25	500±25
	Citation	2.5	2	TBI	<1.25	500±25
	Celebrity	2.5	2	TBI	<1.25	500±25
	Camaro	2.5	2	TBI	<1.25	500±25
	Camaro	5.0	S	CFI	0.525±0.075 400	
	Corvette	5.7	8	CFI	0.525±0.075	400
1984	Camaro	2.5	2	TBI	<1.25	500±25
	Cavalier	2.5	2	TBI	<1.25	500±25
	Celebrity	2.5	R	TBI	<1.25	500±25
	Citation	2.5	R	TBI	<1.25	500±25
	Camaro	5.0	S	CFI	0.525±0.075	400
	Corvette	5.7	8	CFI	0.525±0.075	400
1985	Camaro	2.5	2	TBI	<1.25	500±25
	Camaro	2.8	S	MPI	0.55±0.050	600±50
	Camaro	5.0	F	TBI	0.54±0.075	500±50
	Caprice	4.3	Z	TBI	0.525±0.075	550±50
	Cavalier	2.0	P	TBI	<1.25	650±25
	Cavalier	2.8	W	MPI	0.55±0.050	600±50
	Celebrity	2.5	R	TBI	<1.25	500±25
	Celebrity	2.8	W	MPI	0.55±0.050	600±50
	Citation	2.5	R	TBI	<1.25	500±25
	Citation	2.8	W	MPI	0.55±0.050	600±50
	Corvette	5.7	8	TPI	0.54±0.075	400
	El Camino	4.3	Z	MPI	0.525±0.075	550±50
	Monte Carlo	4.3	Z	MPI	0.525±0.075	550±50
1986	Camaro	2.5	2	TBI	<1.25	500±25
	Camaro	2.8	S	MPI	0.55±0.050	600+50
	Camaro	5.0	F	TPI	0.54±0.075	400
	Caprice	4.3	Z	TBI	<1.25	425±25
	Cavalier	2.0	P	TBI	<1.25	650±25
	Cavalier	2.8	W	MPI	0.55±0.050	600±50
	Celebrity	2.5	R	TBI	<1.25	500±25
	Celebrity	2.8	W	MPI	0.55±0.050	600±50
	Corvette	5.7	Y	TPI	0.54±0.075	400
	El Camino	4.3	Z	TBI	<1.25	425±25
	Monte Carlo	4.3	Z	TBI	<1.25	425±25
1987-88	Beretta	2.0	1	TBI	<1.25	600±25
	Beretta	2.8	W	MPI	0.55±0.10	650
	Camaro	2.8	S	MPI	0.55±0.060	500
	Camaro	5.0	F	TPI	0.54±0.080	400
	Camaro	5.7	8	MPI	0.54±0.080	450
	Caprice	4.3	Z	TBI	<1.25	425±25
	Cavalier	2.0	1	TBI	<1.25	600±25
	Cavalier	2.8	W	MPI	0.55±0.10	650

Year	Model	Engine (liters)	VIN #	Injection system	TPS (volts)	Minimum Air (rpm)
	Celebrity	2.5	R	TBI	<1.25	600±25
	Celebrity	2.8	W	MPI	0.55±0.10	650
	Corsica	2.0	1	TBI	<1.25	600±25
	Corsica	2.8	W	MPI	0.55±0.10	650
	Corvette	5.7	8	TPI	0.54±0.080	450
	El Camino	4.3	Z	TBI	<1.25	425±25
	Monte Carlo	4.3	Z	TBI	<1.25	425±25
1988-90	Beretta	2.0	1	TBI	<1.25	450-650
	Beretta	2.8	W	PFI	0.55+0.10	550+50
	Camaro	2.8	S	PFI	0.55+0.10	IAC 10-15
	Camaro	5.0	E	TBI	<1.25	475+25
	Camaro	5.0	F	TPFI	0.54+0.08	425+25
	Camaro	5.7	8	TPFI	0.54+0.08	425+25
	Caprice	4.3	Z	TPFI	<1.25	425+25
	Cavalier	2.0	W	PFI	<1.25	600+25
	Cavalier	2.8	W	PFI	<1.25	IAC 10-15
	Celebrity	2.5	R	TBI	<1.25	600+50
	Celebrity	2.8	W	PFI	<1.25	IAC 10-15
	Corsica	2.0	1	TBI	<1.25	450-650
	Corsica	2.8	W	PFI	0.55+0.10	550+50
	Lumina	2.5	R	TBI	<1.25	600+50
	Corvette	5.7	8	TPFI	0.54+0.08	425+25
	Monte Carlo	4.3	Z	TBI	<1.25	425+25
	Astro van	2.5	E	TBI	<1.25	450-650
	Lt. truck	2.5	E	TBI	<1.25	450-650
	S-10, S15	2.5	E	TBI	<1.25	450-650
	Astro van	2.5	E	TBI	<1.25	450-650
	Astro Van	2.8	R	TBI	0.48+0.06	450-650
	St. truck	2.8	R	TBI	0.48+0.06	450-650
	S-10, S-15	2.8	R	TBI	0.48+0.06	450-650
	Astro Van	2.8	R	TBI	0.48+0.06	450-650
	Lt. truck	4.3	Z	TBI	<1.25	425+25
	S-10, S-15	4.3	Z	TBI	<1.25	425+25
	Astro van	4.3	Z	TBI	<1.25	425+25
	Lt. Truck	5.0	H	TBI	<1.25	400-600
	Astro van	5.0	H	TBI	<1.25	400-600
	Lt. truck	5.7	K	TBI	<1.25	400-600
	Astro van	5.7	K	TBI	<1.25	400-600
	Lt. truck	7.4	N	TBI	<1.25	400-600

ECM Location

Year	Model	Location
1986-1989	Astro van	Passenger's kick panel
1988-1990	Beretta	Passenger's kick panel
11982-1990	Camaro	Under passenger's dash
1985-1990	Cavalier	Behind glovebox
1985-1988	Celebrity	Behind glovebox
1982-1983	Corvette	Behind driver's seat
1984-1990	Corvette	Under passenger's dash
1984-1990	C/K 10,20,30	Behind glovebox
1982-1988	Monte Carlo	Passenger's kick panel
1984-1990	S-10 Series	Under passenger's dash
1984-1990	Impala/Caprice	Passenger's kick panel

ECM Pinouts

On 1981 and later Chevrolet vehicles, the eighth digit of the VIN (vehicle identification number) identifies the factory-installed engine.

VIN Code E 2.5-Liter Light Truck

A Connectors

Pinout	Wire color	Function
A1	Green and white	Fuel pump relay
A2	Tan and black	Torque converter clutch (AT) or shift light (MT)
A3	Gray	EGR solenoid
A4	Blue	A/C relay
A5	Brown and white	Service Engine Soon light
A6	Pink and black	12 volts, switch ignition
A7	Not used	
A8	Orange	Serial data
A9	White and black	Diagnostic input
A10	Brown	Vehicle speed sensor
A11	Black	Coolant temperature sensor and TPS ground
A12	Black and white	System ground

B Connectors

Pinout	Wire color	Function
B1	Orange	Battery
B2	Tan and white	Fuel pump signal
B3	Black and red	HEI ignition module ground
B4	Not used	
B5	Purple and white	HEI reference
B6	Not used	
B7	Not used	
B8	Dark green and white	A/C signal
B9	Not used	
B10	Orange and black	Park/neutral switch
B11	Not used	
B12	Not used	

C Connectors

Pinout	Wire color	Function
C1	Not used	
C2	Not used	
C3	Green and black	IAC coil B low
C4	Green and white	IAC coil B high
C5	Blue and white	IAC coil A high
C6	Blue and black	IAC coil A low
C7	Not used	
C8	Blue and yellow	Power steering signal
C9	Not used	
C10	Yellow	Coolant temperature sensor
C11	Light green	Manifold absolute pressure
C12	Tan	Manifold air temperature sensor signal
C13	Dark blue	Throttle position sensor
C14	Gray	5-volt reference
C15	Not used	
C16	Orange	Battery

D Connectors

Pinout	Wire color	Function
D1	Tan and white	System ground
D2	Orange and black	MAP ground
D3	Not used	
D4	White	Electronic Spark Timing
D5	Tan and black	Bypass
D6	Tan	Oxygen sensor ground
D7	Purple	Oxygen sensor signal
D8	Not used	
D9	Not used	
D10	Not used	
D11	Not used	
D12	Not used	
D13	Not used	
D14	Not used	
D15	Not used	
D16	Blue	Injector drive

VIN Code F 5.0- and 5.7-Liter TPI Camaro

A Connectors

Pinout	Wire color	Function
A1	Green and white	Fuel pump relay
A2	Dark brown	Air switching control
A3	Green and yellow	Canister purge control
A4	Gray	EGR control
A5	Brown and white	Service Engine Soon light
A6	Pink and black	12 volts, switch ignition
A7	Tan and black	Torque converter clutch or shift light
A8	Orange	Serial data
A9	White and black	Diagnostic input
A10	Brown	Vehicle speed sensor
A11	Black	MAF analog ground
A12	Black and white	System ground

B Connectors

Pinout	Wire color	Function
B1	Orange	Battery
B2	Tan and white	Fuel pump signal
B3	Black and red	HEI ignition module ground
B4	Not used	
B5	Purple and white	HEI reference
B6	Not used	
B7	Black	ESC signal
B8	Green	A/C signal
B9	Not used	
B10	Orange and black	Park/neutral switch
B11	Dark green	Mass airflow sensor signal
B12	Not used	

C Connectors

Pinout	Wire color	Function
C1	Dark green and white	Fan relay control
C2	Black and pink	Converter air control
C3	Green and black	IAC coil B low
C4	Green and white	IAC coil B high
C5	Blue and white	IAC coil A high
C6	Blue and black	IAC coil A low
C7	Not used	
C8	Not used	
C9	Not used	
C10	Yellow	Coolant temperature sensor
C11	Not used	
C12	Tan	Manifold air temperature signal
C13	Dark blue	Throttle position sensor
C14	Gray	TPS 5-volt reference
C15	Light green	EGR diagnostic switch
C16	Orange	Battery 12 volts

D Connectors

Pinout	Wire color	Function
D1	Black and white	System ground
D2	Black	TPS, CTS, MAT 5 volt return
D3	Black and white	System ground
D4	White	EST control
D5	Tan and black	Bypass
D6	Tan	Oxygen sensor ground
D7	Purple	Oxygen sensor signal
D8	Not used	
D9	Not used	
D10	Black and white	System ground
D11	Dark blue	A/C pressure fan switch
D12	Black	Burn-off relay control
D13	Not used	
D14	Not used	
D15	Light blue	Injector drive 1, 3, 5, 7
D16	Light blue	Injector drive 2, 4, 6, 8

VIN Code P 2.0-Liter

A Connectors

Pinout	Wire color	Function
A1	Green and white	Fuel pump relay
A2	Not used	
A3	Light blue	A/C relay
A4	Not used	
A5	Brown and white	Service Engine Soon light
A6	Pink and black	12 volts, switch ignition
A7	Tan and black	Torque converter clutch or shift light
A8	Orange	Serial data
A9	White and black	Diagnostic input
A10	Brown	Vehicle speed sensor
A11	Black	Coolant temperature sensor and TPS ground
A12	Black and white	System ground

B Connectors

Pinout	Wire color	Function
B1	Orange	Battery
B2	Tan and white	Voltage monitor
B3	Black and red	HEI ignition module
B4	Not used	
B5	Purple and white	HEI reference
B6	Not used	
B7	Not used	
B8	Dark green	A/C signal
B9	Black	System ground
B10	Orange and black	Park/neutral switch
B11	Not used	
B12	Not used	

C Connectors

Pinout	Wire color	Function
C1	Green and white	Cooling fan control
C2	Not used	
C3	Green and black	IAC coil B low
C4	Green and white	IAC coil B high
C5	Blue and white	IAC coil A high
C6	Blue and black	IAC coil A low
C7	Not used	
C8	Gray	Power steering signal
C9	Blue and white	Crank input
C10	Yellow	Coolant temperature sensor
C11	Light green	Manifold absolute pressure
C12	Not used	
C13	Dark blue	Throttle position sensor
C14	Gray	5-volt reference
C15	Not used	
C16	Orange	Battery

D Connectors

Pinout	Wire color	Function
D1	Tan and white	System ground
D2	Black	MAP ground
D3	Not used	
D4	White	Electronic Spark Timing
D5	Tan and black	Bypass
D6	Tan	Oxygen sensor ground
D7	Purple	Oxygen sensor signal
D8	Not used	
D9	Black	Jumper, manual transmission only
D10	Black	Jumper, manual transmission only
D11	Not used	
D12	Not used	
D13	Not used	
D14	Not used	
D15	Not used	
D16	Light blue	Injector drive

VIN Code R 2.5-Liter

White Collectors

Pinout	Wire color	Function
1	Not used	
2	Brown	Vehicle speed sensor
3	White and black	Diagnostic test
4	Not used	
5	Orange and black	Park-neutral switch
6	Not used	
7	Orange	Serial data
8	Blue	Injector drive
9	Not used	
10	Orange	Battery
11	Gray	5-volt reference
12	Black and white	ECM ground
13	Black and white	ECM ground
14	Black and orange	MAP ground
15	Orange	Battery
16	Pink and black	12 volts, switched ignition
17	Not used	
18	Dark green and white	Fuel-pump relay drive
19	Tan and black	Torque converter clutch or shift light
20	Brown and white	Service Engine Soon light
21	Dark green and white	A/C clutch
22	Not used	
23	Not used	
24	Not used	

Black Connectors

Pinout	Wire color	Function
1	Yellow	Crank
2	Purple and white	HEI reference
3	Black and red	HEI ignition distributor ground
4	Yellow	Coolant sensor signal
5	Dark blue	TPS signal
6	Light blue and orange	Power steering signal
7	Dark blue	A/C clutch relay
8	Purple	Oxygen sensor signal
9	Light blue and black	Cooling fan
10	Tan and black	EST by-pass
11	Black	Coolant and TPS ground
12	Light green and white	IAC coil B high
13	Light green and black	IAC coil B low
14	Not used	
15	Tan	Oxygen sensor ground
16	Not used	
17	Light blue and white	IAC coil A high
18	Light blue and black	IAC coil A low
19	White	EST control
20	Light green	Manifold pressure signal
21	Not used	
22	Not used	

VIN Code R 2.8-Liter

A Connectors

Pinout	Wire color	Function
A1	Green and white	Fuel pump relay
A2	Dark brown	A/C clutch control
A3	Not used	
A4	Gray	EGR control
A5	Brown and white	Service Engine Soon light
A6	Pink and black	12 volts, switch ignition
A7	Tan and black	Torque converter clutch or shift light
A8	Orange and green	Serial data
A9	White and black	Diagnostic input
A10	Brown	Vehicle speed sensor
A11	Purple	MAP ground
A12	Black and white	System ground

B Connectors

Pinout	Wire color	Function
B1	Orange	Battery
B2	Tan and white	Fuel pump signal
B3	Black and red	HEI ignition module ground
B4	Not used	
B5	Purple and white	HEI reference
B6	Not used	
B7	Black	ESC signal
B8	Dark green	A/C signal
B9	Brown	EGR switch signal
B10	Orange and black	Park-neutral switch
B11	Not used	
B12	Not used	

C Connectors

Pinout	Wire color	Function
C1	Pink and black	EAC solenoid (Federal)
C2	Brown	EAC solenoid (California)
C3	Green and black	IAC coil B low
C4	Green and white	IAC coil B high
C5	Blue and white	IAC coil A high
C6	Blue and black	IAC coil A low
C7	Not used	
C8	Not used	
C9	Purple and white	Crank input
C10	Yellow	Coolant temperature sensor
C11	Light green	Manifold absolute pressure
C12	Dark blue and white	Elapsed timer module
C13	Dark blue	Throttle position sensor
C14	Gray	5-volt reference
C15	Not used	
C16	Orange	Battery 12 volts

D Connectors

Pinout	Wire color	Function
D1	Tan and white	System ground
D2	Black	MAP ground
D3	Not used	
D4	White	Electronic Spark Timing
D5	Tan and black	Bypass
D6	Tan	Oxygen sensor ground
D7	Purple	Oxygen sensor signal
D8	Not used	
D9	Not used	
D10	Not used	
D11	Not used	
D12	Not used	
D13	Not used	
D14	Light green	Injector drive B
D15	Not used	
D16	Light blue	Injector drive A

VIN Code R Lumina

A Connectors (Black)

Pinout	Wire color	Function
A1	Light blue and white	IAC A high
A2	Light green and black	IAC B low
A3	Not used	
A4	Not used	
A5	Not used	
A6	Not used	
A7	Light blue and black	IAC A low
A8	Light green and white	IAC B high
A9	Dark green and white	Cooling fan
A10	Not used	
A11	Not used	
A12	Not used	
A13	Not used	
A14	Not used	
A15	Not used	
A16	Purple	Oxygen sensor signal
A17	Not used	
A18	Not used	
A19	Not used	
A20	Not used	
A21	Not used	
A22	Tan	Oxygen sensor ground

B Connectors (Natural)

Pinout	Wire color	Function
B1	Brown and white	Service Engine Soon light
B2	Not used	
B3	White and black	Diagnostic input
B4	Not used	
B5	Orange	Serial data
B6	Not used	
B7	Tan and black	Torque converter clutch control
B8	Dark green	Vehicle speed sensor output
B9	Not used	
B10	Pink and black	Switched ignition voltage
B11	Not used	
B12	Dark blue	A/C low-pressure input
B13	Not used	
B14	Not used	
B15	Not used	
B16	Not used	
B17	Not used	
B18	Not used	
B19	Not used	
B20	Not used	
B21	Not used	
B22	Not used	

C Connectors (Gray)

Pinout	Wire color	Function
C1	Orange	Battery 12 volts
C2	Purple	Vehicle speed sensor input
C3	Tan and black	Ignition by-pass
C4	Tan	Manifold air temperature sensor
C5	Black	MAP, MAT ground
C6	Black and white	System ground
C7	Gray	MAP 5-volt reference
C8	Yellow	Vehicle speed sensor signal low
C9	White	EST control
C10	Black	TPS, CTS ground
C11	Not used	
C12	Gray	TPS 5-volt reference
C13	Dark green and yellow	A/C relay control
C14	Not used	
C15	Dark blue	TPS signal
C16	Yellow	Coolant temperature signal
C17	Light blue	A/C request
C18	Not used	
C19	Not used	
C20	Not used	
C21	Not used	
C22	Light green	MAP signal

D Connectors (Blue)

Pinout	Wire color	Function
D1	Not used	
D2	Not used	
D3	Light blue	Injector driver
D4	Tan	Peak and hold jumper A
D5	Not used	
D6	Not used	
D7	Dark green and white	Fuel-pump relay
D8	Tan	Peak and hold jumper A
D9	Light blue	Injector drive
D10	Not used	
D11	Orange and black	Park-neutral switch
D12	Tan and white	System ground
D13	Purple and white	EST
D14	Not used	
D15	Not used	
D16	Light blue and orange	Power-steering pressure switch
D17	Orange	Battery 12 volts
D18	Not used	
D19	Black and red	EST (ignition ground)
D20	Not used	
D21	Not used	
D22	Not used	

VIN Code S 5.0-Liter

White Connectors

Pinout	Wire color	Function
1	Not used	
2	Brown	Vehicle speed sensor
3	White and black	Diagnostic test
4	Black	ESC (Electronic Spark Control)
5	Orange and black	Park-neutral switch
6	Light blue	Dual-injector select
7	Orange	Serial data
8	Blue	Rear-injector drive
9	Light green	Front-injector drive
10	Orange	Battery
11	Gray	5-volt reference
12	Black and white	ECM ground
13	Black and white	ECM ground
14	Black and white	MAP ground
15	Orange	Battery
16	Pink and black	12 volts, switch ignition
17	Tan and white	Voltage monitor
18	Dark green and white	Fuel-pump relay drive
19	Tan and black	Torque converter clutch or shift light
20	Brown and white	Service Engine Soon light
21	Dark green and white	A/C clutch
22	Dark green	Fourth gear signal (manual transmission)
23	Dark blue and white	Cold-start module
24	Not used	

Black Connectors

Pinout	Wire color	Function
1	Purple	Crank
2	Purple and white	HEI reference
3	Black and red	HEI (ignition) distributor ground
4	Yellow	Coolant sensor signal
5	Dark blue	TPS signal
6	Gray and red	Fourth gear automatic transmission, second gear manual transmission
7	Not used	
8	Purple	Oxygen sensor signal
9	Dark green and yellow	Canister purge
10	Tan and black	EST by-pass
11	Black	Coolant and TPS ground
12	Light green and white	IAC coil B high
13	Light green and black	IAC coil B low
14	Brown	Air switch solenoid
15	Tan	Oxygen sensor ground
16	Black and pink	Air diverter solenoid
17	Light blue and red	IAC coil A high
18	Light blue and red	IAC coil A low
19	White	EST control
20	Light green	Manifold pressure signal
21	Not used	
22	Gray	EGR

VIN Code T 3.1-Liter Lumina

A Connectors

Pinout	Wire color	Function
A1	Light blue and white	IAC coil A high
A2	Light green and black	IAC coil B low
A3	Dark blue and white	Fan #2 control
A4	Gray and red	EGR control
A5	Not used	
A6	Not used	
A7	Light blue and black	IAC coil A low
A8	Light green and white	IAC coil B high
A9	Dark green and white	Fan #1 control
A10	Dark green and yellow	Canister purge
A11	Dark blue	ESC signal
A12	Dark green and yellow	A/C relay control
A13	Yellow and black	Check coolant light
A14	Not used	
A15	Not used	
A16	Purple	Oxygen sensor signal
A17	Not used	
A18	Not used	
A19	Not used	
A20	Gray	Fuel-pump signal
A21	Not used	
A22	Tan	Oxygen sensor ground

B Connectors

Pinout	Wire color	Function
B1	Brown and white	Service Engine Soon light
B2	Not used	
B3	White and black	Diagnostic test
B4	Not used	
B5	Orange	Serial data in
B6	Not used	
B7	Tan and black	Torque converter clutch or shift light
B8	Dark green	Vehicle speed out
B9	Not used	
B10	Pink and black	Ignition feed
B11	Not used	
B12	Dark blue	A/C low-pressure signal
B13	Not used	
B14	Not used	
B15	Not used	
B16	Not used	
B17	Not used	
B18	Not used	
B19	Not used	
B20	Light green and black	Fan #2 request
B21	Not used	
B22	Not used	

C Connectors

Pinout	Wire color	Function
C1	Orange	Battery feed
C2	Purple	Magnetic vehicle speed sensor low
C3	Tan and black	DIS by-pass
C4	Tan	Manifold air temperature sensor signal
C5	Black	MAP, CTS ground
C6	Black and white	EGR ground
C7	Gray	Coolant level sensor
C8	Yellow	Magnetic vehicle speed sensor high
C9	White	EST control
C10	Black	Manifold air temperature sensor ground
C11	Not used	
C12	Gray	EGR, TPS, MAP 5-volt reference
C13	Not used	
C14	Gray and red	Coolant level signal
C15	Dark blue	TPS signal
C16	Yellow	Coolant temperature signal
C17	Light blue	A/C request
C18	Red	EGR position signal
C19	Not used	
C20	White	Second gear switch (if used)
C21	Not used	
C22	Light green	MAP signal

D Connectors

Pinout	Wire color	Function
D1	Not used	
D2	Not used	
D3	Light green	Injector drive 1, 3, 5
D4	Black and white	Ground
D5	Not used	
D6	Dark green	Third gear signal
D7	Dark green and white	Fuel-pump relay drive
D8	Not used	
D9	Light blue	Injector drive 2, 4, 6
D10	Tan and white	Ground
D11	Orange and black	Park-neutral switch
D12	Tan and white	Ground
D13	Purple and white	DIS reference
D14	Not used	
D15	Not used	
D16	Light blue and orange	Power-steering pressure signal
D17	Orange	Battery feed
D18	Not used	
D19	Black and red	Ignition ground
D20	Not used	
D21	Dark green	Fan #1 request
D22	Light blue	Fourth gear signal

VIN Code W 2.8-Liter Cavalier

A Connectors

Pinout	Wire color	Function
A1	Not used	
A2	Not used	
A3	Red	EGR position
A4	Gray and red	5-volt reference
A5	Gray	5-volt reference
A6	Pink and black	Switched ignition voltage
A7	Not used	
A8	Not used	
A9	Orange	Serial data
A10	Not used	
A11	Dark green and white	Fuel-pump relay control
A12	Black	Power ground

B Connectors

Pinout	Wire color	Function
B1	Orange	Battery
B2	Not used	
B3	Not used	
B4	Not used	
B5	Black	Sensor ground
B6	Purple	Sensor ground
B7	Not used	
B8	Not used	
B9	Purple	Vehicle speed sensor low
B10	Yellow	Vehicle speed sensor high
B11	Dark green	To instrument panel (4,000 pulses per mile)
B12	Not used	

C Connectors

Pinout	Wire color	Function
C1	Red	To cruise control (2,000 pulses per mile)
C2	Not used	
C3	Not used	
C4	Purple	Battery voltage (if used)
C5	Not used	
C6	Not used	
C7	Tan and black	By-pass
C8	White	EST
C9	Light green	A/C request
C10	Not used	
C11	Light blue	Injector drive 2, 4, 6
C12	Light green	Injector drive 1, 3, 5
C13	Not used	
C14	Not used	
C15	Not used	
C16	Orange	Battery

D Connectors

Pinout	Wire color	Function
D1	Tan and white	Power ground
D2	Not used	
D3	Not used	
D4	Not used	
D5	Not used	
D6	Black and white	Injector ground
D7	Black and white	Injector ground
D8	Purple and white	HEI reference
D9	Black and red	Ignition ground
D10	Not used	
D11	Not used	
D12	Dark green and white	A/C pressure fan switch
D13	Light blue and orange	Power-steering pressure switch
D14	Not used	
D15	Not used	
D16	Orange and black	Park-neutral switch

E Connectors

Pinout	Wire color	Function
E1	Not used	
E2	Not used	
E3	Light blue and white	IAC coil A high
E4	Light blue and black	IAC coil A low
E5	Light green and white	IAC coil B high
E6	Light green and black	IAC coil B low
E7	Brown and white	Service Engine Soon light
E8	Dark green and white	Fan relay control
E9	Gray	EGR control
E10	Not used	
E11	Not used	
E12	White and black	Diagnostic terminal
E13	Gray	Fuel pump signal
E14	Purple	Oxygen sensor signal
E15	Tan	Oxygen sensor ground
E16	Yellow	Coolant sensor signal

F Connectors

Pinout	Wire color	Function
F1	Green and white	A/C Control relay
F2	Not used	
F3	Not used	
F4	Not used	
F5	Not used	
F6	Tan and black	Torque converter clutch control
F7	Dark green and yellow	Purge control
F8	Not used	
F9	Dark blue	ESC signal
F10	Not used	
F11	Not used	
F12	Not used	
F13	Dark blue	TPS signal
F14	Not used	
F15	Light green	MAP signal
F16	Tan	MAT signal

VIN Code W 2.8-Liter Corsica and Beretta

A Connectors

Pinout	Wire color	Function
A1	Not used	
A2	Not used	
A3	Pink	EGR position
A4	Gray and red	5-volt reference
A5	Gray	5-volt reference
A6	Pink and black	Switched ignition voltage
A7	Not used	
A8	Not used	
A9	Orange	Serial data
A10	Not used	
A11	Dark green	Fuel-pump relay control
A12	Black and white	Power ground

B Connectors

Pinout	Wire color	Function
B1	Orange	Battery
B2	Not used	
B3	Not used	
B4	Not used	
B5	Black	Sensor ground
B6	Purple and yellow	Sensor ground
B7	Not used	
B8	Not used	
B9	Purple	Vehicle speed sensor low
B10	Yellow	Vehicle speed sensor high
B11	Gray and black	To instrument panel (4,000 pulses per mile)
B12	Not used	

C Connectors

Pinout	Wire color	Function
C1	Not used	
C2	Not used	
C3	Not used	
C4	Not used	
C5	Not used	
C6	Not used	
C7	Tan and black	By-pass
C8	White	EST
C9	Black and pink, Black and green	With A/C on, A/C request
C10	Not used	
C11	Light blue	Injector drive 2, 4, 6
C12	Light green	Injector drive 1, 3, 5
C13	Not used	
C14	Not used	
C15	Not used	
C16	Orange	Battery

D Connectors

Pinout	Wire color	Function
D1	Tan and white	Power ground
D2	Not used	
D3	Not used	
D4	Not used	
D5	Not used	
D6	Black and white	Injector ground
D7	Black and white	Injector ground
D8	Purple and white	HEI reference
D9	Black and red	Ignition ground
D10	Not used	
D11	Not used	
D12	Black and yellow	A/C pressure fan switch
D13	Tan	Power-steering pressure switch
D14	Not used	
D15	Not used	
D16	Orange and black	Park-neutral switch

E Connectors

Pinout	Wire color	Function
E1	Not used	
E2	Not used	
E3	Dark green	IAC coil A high
E4	Green and white	IAC coil A low
E5	Dark blue	IAC coil B high
E6	Blue and white	IAC coil B low
E7	Brown and white	Service Engine Soon light
E8	Dark green and white	Fan relay control
E9	Gray	EGR control
E10	Not used	
E11	Not used	
E12	White and black	Diagnostic terminal
E13	Tan and white	Fuel pump signal
E14	Purple	Oxygen sensor signal
E15	Tan	Oxygen sensor signal
E16	Gray	Coolant Sensor signal

F Connectors

Pinout	Wire color	Function
F1	Brown	A/C control relay
F2	Not used	
F3	Not used	
F4	Not used	
F5	Not used	
F6	Tan and black	Torque converter clutch control
F7	Dark green and yellow	Purge control
F8	Not used	
F9	Dark blue	ESC signal
F10	Not used	
F11	Not used	
F12	Not used	
F13	Dark blue	TPS signal
F14	Not used	
F15	Purple and white	MAP signal
F16	Black and pink	MAT signal

VIN Code Z 4.3-Liter

A Connectors

Pinout	Wire color	Function
A1	Green and white	Fuel-pump relay
A2	Not used	
A3	Green and yellow	Canister purge control
A4	Gray and red	EGR control
A5	Brown and white	Service Engine Soon light
A6	Pink and black	12 volts, switch ignition
A7	Tan and black	Torque converter clutch or shift light
A8	Orange	Serial data
A9	White and black	Diagnostic input
A10	Brown	Vehicle speed sensor
A11	Black	Coolant temperature sensor and TPS ground
A12	Black and white	System ground

B Connectors

Pinout	Wire color	Function
B1	Orange	Battery
B2	Brown	Fuel-pump signal
B3	Black and red	HEI ignition module ground
B4	Not used	
B5	Purple and white	HEI reference
B6	Not used	
B7	Black	ESC signal
B8	Dark green	A/C signal
B9	Blue	EGR temperature sensor
B10	Orange and black	Park-neutral switch
B11	Not used	
B12	Not used	

C Connectors

Pinout	Wire color	Function
C1	Black and pink	AIR divert solenoid
C2	Brown	AIR switch solenoid
C3	Green and black	IAC coil B low
C4	Green and white	IAC coil B high
C5	Blue and white	IAC coil A high
C6	Blue and black	IAC coil A low
C7	Light blue	Fourth gear
C8	Not used	
C9	Purple and white	Crank input
C10	Yellow	Coolant temperature sensor
C11	Light green	Manifold absolute pressure
C12	Not used	
C13	Dark blue	Throttle position sensor
C14	Gray	5-volt reference
C15	Light green	Injector B
C16	Orange	Battery

D Connectors

Pinout	Wire color	Function
D1	Black and white	System ground
D2	Purple	MAP ground
D3	Not used	
D4	White	Electronic Spark Timing
D5	Tan and black	Bypass
D6	Tan	Oxygen sensor ground
D7	Purple	Oxygen sensor signal
D8	Not used	
D9	Not used	
D10	Not used	
D11	Not used	
D12	Not used	
D13	Not used	
D14	Light green	Injector B
D15	Light blue	Injector A
D16	Light blue	Injector A

VIN Code 1 2.0-Liter Cavalier

W Connectors

Pinout	Wire color	Function
W1	Blue	Injector drive
W2	Dark blue and white	A/C request
W3	Red	Vehicle speed sensor (2,000 pulses per mile)
W4	Gray and black	Cruise
W5	Light blue and orange	Power steering signal
W6	White	EST
W7	Light blue and white	IAC coil A high
W8	Light green and black	IAC coil B low
W9	Light green and white	IAC coil B high
W10	Orange	Battery
W11	Gray	5-volt reference
W12	Black and white	ECM ground
W13	Tan and white	ECM ground
W14	Black and orange	MAP, MAT ground
W15	Orange	Battery
W16	Pink and black	Switched ignition voltage
W17	Light blue and black	IAC coil A low
W18	Orange and black	Park-neutral switch
W19	Tan and black	Bypass
W20	Gray	Cruise
W21	Dark blue	Cruise
W22	Dark green	Vehicle speed sensor output
W23	Not used	
W24	Dark green and white	Fuel pump

B Connectors

Pinout	Wire color	Function
B1	Orange	Serial data
B2	Purple	Oxygen sensor signal
B3	Dark green and white	A/C clutch relay
B4	Not used	
B5	Tan	Manifold air temperature sensor
B6	Yellow	Vehicle speed sensor input
B7	Tan and black	Torque converter clutch
B8	Yellow	Coolant temperature sensor
B9	Purple and white	HEI reference
B10	Light green	Cruise
B11	Dark blue and white	Cruise
B12	Black	Coolant and TPS ground
B13	Purple	Vehicle speed sensor input
B14	Dark brown	Cruise
B15	White and black	ALDL diagnostic
B16	Black and red	Ignition ground
B17	Not used	
B18	White	Cruise
B19	Dark blue	TPS signal
B20	Light green	MAP signal
B21	Dark green and white	Engine cooling fan
B22	Brown and white	Service Engine Soon light
B23	Tan	Oxygen sensor ground
B24	Not used	

VIN Code 1 Corsica and Beretta

W Connectors

Pinout	Wire color	Function
W1	Blue	Injector drive
W2	Light green	A/C request
W3	Not used	
W4	Gray and black	Cruise
W5	Tan	Power steering signal
W6	White	EST
W7	Dark green	IAC coil A high
W8	Dark blue and white	IAC coil B low
W9	Dark blue	IAC coil B high
W10	Orange	Battery
W11	Gray	5-volt reference
W12	Tan and white	ECM ground
W13	Black and white	ECM ground
W14	Black and orange	MAP, MAT ground
W15	Orange	Battery
W16	Pink and black	Switched ignition voltage
W17	Dark green and white	IAC coil A low
W18	Orange and black	Park-neutral switch
W19	Tan and black	Bypass
W20	Gray	Cruise
W21	Dark blue	Cruise
W22	Gray and black	Vehicle speed sensor output
W23	Not used	
W24	Dark green	Fuel pump

B Connectors

Pinout	Wire color	Function
B1	Orange	Serial data
B2	Purple	Oxygen sensor signal
B3	Brown	A/C clutch relay
B4	Not used	
B5	Black and pink	Manifold air temperature sensor
B6	Yellow	Vehicle speed sensor input
B7	Tan and black	Torque converter clutch
B8	Gray	Coolant temperature sensor
B9	Purple and white	HEI reference
B10	Light green	Cruise
B11	Dark blue and white	Cruise
B12	Purple and yellow	Coolant and TPS ground
B13	Purple	Vehicle speed sensor input
B14	Dark brown	Cruise
B15	White and black	ALDL diagnostic
B16	Black and red	Ignition ground
B17	Not used	
B18	Not used	
B19	Dark blue	TPS signal
B20	Purple and white	MAP signal
B21	Dark green and white	Engine cooling fan
B22	Brown and white	Service Engine Soon light
B23	Tan	Oxygen sensor ground
B24	Not used	

VIN Code 8 5.7-Liter Corvette CFI

White Connectors

Pinout	Wire color	Function
1	Not used	
2	Brown	Vehicle speed sensor
3	White and black	Diagnostic test
4	Black	ESC
5	Orange and black	Park-neutral switch
6	Light blue	Dual-injector select
7	Orange	Serial data
8	Blue	Rear-injector drive
9	Light green	Front-injector drive
10	Orange	Battery
11	Gray	5-volt reference
12	Black and white	ECM ground
13	Black and white	ECM ground
14	Black and white	MAP ground
15	Orange	Battery
16	Pink and black	12 volts, switched ignition
17	Tan and white	Voltage monitor
18	Dark green and white	Fuel-pump relay drive
19	Tan and black	Torque converter clutch or shift light
20	Brown and white	Service Engine Soon light
21	Dark green and white	A/C clutch
22	Dark green	Fourth gear signal (manual transmission)
23	Dark blue and white	Cold Start Module
24	Not used	

Black Connectors

Pinout	Wirecolor	Function
1	Purple	Crank
2	Purple and white	HEI reference
3	Black and red	HEI ignition module distributor ground
4	Yellow	Coolant sensor signal
5	Dark blue	TPS signal
6	Gray and red	Fourth gear automatic transmission, second gear manual transmission
7	Not used	
8	Purple	Oxygen sensor signal
9	Dark green and yellow	Canister purge
10	Tan and black	EST bypass
11	Black	Coolant and TPS ground
12	Light green and white	IAC coil B high
13	Light green and black	IAC coil B low
14	Brown	Air switch solenoid
15	Tan	Oxygen sensor ground
16	Black and pink	Air diverter solenoid
17	Light blue and red	IAC coil A high
18	Light blue and red	IAC coil A low
19	White	EST control
20	Light green	Manifold pressure signal
21	Not used	
22	Gray	EGR

VIN Code 8 Corvette 5.7-Liter TPI

A Connectors

Pinout	Wire color	Function
A1	Green and white	Fuel-pump relay
A2	Dark brown	Airswitching control
A3	Green and yellow	Canister purge control
A4	Gray	EGR control
A5	Brown and white	Service Engine Soon light
A6	Pink and black	12 volts, switch ignition
A7	Tan and black	Torque converter clutch or shift light
A8	Orange	Serial data
A9	White and black	Diagnostic input
A10	Brown	Vehicle speed sensor
A11	Dark green	Manifold air temperature signal
A12	Black and white	System ground

B Connectors

Pinout	Wire color	Function
B1	Orange	Battery
B2	Red	Fuel pump signal
B3	Black and red	HEI ignition module ground
B4	White	EST control
B5	Purple and white	HEI reference
B6	Not used	
B7	Black	ESC signal
B8	Green and white	A/C signal
B9	Not used	
B10	Orange and black	Park-neutral switch
B11	Dark green	Mass air flow sensor signal
B12	Not used	

C Connectors

Pinout	Wire color	Function
C1	Not used	
C2	Black and pink	Air control
C3	Green and black	IAC coil B low
C4	Green and white	IAC coil B high
C5	Blue and white	IAC coil A high
C6	Blue and black	IAC coil A
C7	Black and blue	Overdrive request
C8	Blue	Fourth gear automatic transmission, first gear manual transmission
C9	Dark green	Fan request
D10	Yellow	Coolant temperature sensor
D11	Not used	
C12	Dark blue	Back-up mode TPS signal
C13	Dark blue	Throttle position sensor
C14	Gray	TPS 5-volt reference
C15	Light green	Injector drive 2, 4, 6, 8 B
C16	Orange	Battery 12 volts

D Connectors

Pinout	Wire color	Function
D1	Black and white	System ground
D2	Green and white	Cooling fan control
D3	Black and white	Cylinder select eight-cylinder ground
D4	Not used	
D5	Tan and black	By-pass
D6	Tan	Oxygen sensor ground
D7	Purple	Oxygen sensor signal
D8	Green	EGR diagnostic switch
D9	Not used	
D10	Black and white	MAF ground
D11	Not used	
D12	Black	TPS ground
D13	Black	Coolant, MAF, MAT sensor ground
D14	Light green	Injector drive 2, 4, 6, 8 B
D15	Light blue	Injector drive 1, 3, 7 A
D16	Light blue	Injector drive 1, 3, 5, 7 A

SOURCES

Airsensors Inc.
708 Industry Drive
Tukwilla, WA 98188

B&M Automotive Products
9152 Independence Avenue
Chatsworth, CA 91311

Competition Cams
3406 Democrat Road
Memphis, TN 38118

Crane Cams
530 Fentress Boulevard
Daytona Beach, FL 32014

Digital Fuel Injection
37742 Hills Tech Drive
Farmington Hills, MI 48024

Edelbrock Corp.
2700 California Street
Torrance, CA 90503

Gale Banks Engineering
546 Duggan Avenue
Azusa, CA 91702

Holley Replacement Parts Division
11955 E. Nine Mile Road
Warren, MI 48089

Hypertech, Inc.
2104 Hillshire
Memphis, TN 38133

Itac Automotive
3121 Benton Drive
Garland, TX 75042

MSD/Autotronic Controls Corp.
1490 Henry Brennan Drive
El Paso, TX 79936

OTC
655 Eisenhower Drive
Owatonna, MN 55060

Paxton Superchargers
929 Olympic Boulevard
Santa Monica, CA 90404

T.P.I. Specialties, Inc.
4255 County Road 10 East
Chaska, MN 55318

Turbo City
1137 W. Katella Avenue
Orange, CA 92667

Turbo Performance Products
7355 Varna Avenue
North Hollywood, CA 91605

Veloce Distributing (Micro Dynamcis)
5003 South Genesee Street
Seattle, WA 98118

INDEX

AC rotational pulse generator, 23
AC voltmeter, 29
Air change temperature sensor, 43
Air conditioner compressor clutch, 53
Air filter, 49, 87, 142
Air management system, 53
Air pump system, 53
Air pump, 74, 89, 91, 147
Air-fuel ratio, 13, 34, 39, 42, 45, 47, 48, 55, 57, 59, 61, 69, 150, 161
Airflow rate, 102
ALCL connector, 90
Amplitude, 16
Analog tachometer, 29
Analog voltmeter, 28
Bendix Electrojector, 7
Block learn multiplier, 46, 47, 74, 94, 103
Bosch D-Jetronic, 7, 9
Bosch MAF sensor, 41
Bureau of Auto Repair, 142, 154
Cadillac-Bendix system, 9
California Air Resources Board, 153, 154
Cam sensor, 78, 80
Catalytic converter, 59
Check Engine light, 90–92, 109, 111, 119, 121–123, 126–128, 130, 133–136
Clear flood program, 48
Code 12, 92, 159
Code 13, 92, 118
Code 14, 92, 100, 121, 122, 139
Code 15, 92, 122, 139
Code 16, 93
Code 21, 93, 123
Code 22, 93, 123
Code 23, 93, 100, 126–128, 131, 139
Code 24, 93, 100, 128–130
Code 25, 94, 131, 139
Code 26, 131
Code 32, 94, 105, 131, 133, 134
Code 33, 95, 98, 102, 103, 134–136
Code 34, 95, 98, 125, 135, 136

Code 35, 136
Code 36, 96, 137
Code 41, 96, 137
Code 42, 96, 137
Code 43, 96, 138
Code 44, 97, 104, 138
Code 45, 97, 138
Code 46, 97
Code 51, 97, 139
Code 52, 97, 139
Code 53, 97
Code 54, 97, 105
Code 55, 98, 139
Code 63, 98
Code 64, 98
Cold-start injector, 38
Compression test, 76, 85
Coolant temperature override switch, 50
Coolant temperature sensor, 39, 43, 47, 50, 52, 53
Corporate average fuel economy, 4
Crank sensor, 78, 79
Crosscount readings, 74
Crossfire Injection, 10, 36, 71
DC frequency pulse generator, 22, 23
DC rotational pulse generators, 23
Delco 2 MAF sensor, 43
Delco MAF sensor, 43
Diagnostic scanner, 31
Digital fuel injection, 9, 157
Digital logic probe, 30
Digital tachometer, 29
Digital voltmeter, 28, 29
Direct Ignition System, 23, 79
Direct injection, 7
Distributor cap, 61, 63, 72, 73, 79, 81
Ducting, 49
Duty cycle, 16–18, 30, 52
Dwell meter, 29
EGR valve, 89, 133, 134
Electrochemical method, 12

Electronic control module, 12, 13, 15, 21, 35, 45, 65
Electronic control unit, 156
Electronic spark control, 152
Electronic spark timing, 50
Engine control module, 159
Environment Protection Agency, 4, 11, 75, 139
Evaporative control system, 59
Exhaust gas recirculation, 51, 58
Frequency, 16
Fuel filter, 36, 37, 81, 161
Fuel pump, 35, 36, 39, 46, 86, 105, 109, 156, 161, 163
Fuel rail, 37, 39
Fuel tank, 35, 36
Fuel-pressure gauge, 30
Fuel-pressure regulator, 36, 37, 38, 39
Greenhouse effect, 55
Hall effects switch, 24
Hall effects, 16
High energy ignition, 29
Hitachi MAF sensor, 43
Idle air control, 49, 52, 70, 77, 87, 99, 101
Ignition coil, 13, 17, 60, 61
Ignition module, 65, 66, 77–80, 96
Ignition timing, 65
Indianapolis 500, 7
Indirect injection, 7
Induction method, 13
Intake manifold, 49
Integrator, 104, 105, 170
Kirchoff's Law, 15
Knock sensor, 50, 170
Light emitting diode, 20
Manifold vacuum, 40, 50, 59
MAP sensor, 22, 40, 81
Normally-grounded circuit, 24
Normally-powered circuit, 24
Offenhauser, 7

Ohm's Law, 15
Ohmmeter, 29
On-Board Diagnostics System Generation
    2, 11, 158
Oscilloscope, 16, 30, 63, 83
Oxygen sensor, 12, 24, 29, 44, 47, 53, 59,
    68, 69, 72–74, 87, 90–92, 97, 101,
    104, 118–120, 138, 152, 169
PCV valve, 88
Port fuel injection, 10, 35, 38, 43, 71
Positive crankcase ventilation, 58
Potentiometer, 20
Programmable read only memory, 21, 40,
    46, 153
Pulse width, 17
Radio frequencies, 21
Random access memory, 21, 46
Read only memory, 21, 45
Resistor, 19, 20, 22

Roadshock simulator, 30
Robert Bosch Company, 7
Rochester Ramjet, 7
S1 Sensor Circuit, 112
S2 Sensor Circuit, 112
S3 and S4 Sensor Circuits, 113
S5 Sensor Circuit, 113
S6 Sensor Circuit, 114
S7 Sensor Circuit, 114
S8, Sensor Circuit, 115
Sine wave, 16
Spark initiation voltage, 61
Spark plug wires, 63, 83
Spark plugs, 60–62, 75, 77, 80, 81, 83
Square wave, 16
Static electricity, 13
Strain gauge, 20
Switch to pull low circuit, 21
Switch-to-voltage circuit, 21

Test light, 29, 30
Thermistor, 19, 41
Thermo-time switch, 38
Three-wire-variable-voltage circuit, 22
Throttle body injection, 7, 10, 35, 38, 83,
    156
Throttle plates, 70, 87
Throttle position sensor, 22, 28, 45, 47,
    49, 70, 82, 83, 93, 100, 169
Transistor, 20
Tuned port injection, 10, 35, 41, 49, 53,
    95, 141, 142, 151, 157
Vacuum gauge, 30
Vacuum pump, 30
Variable-resistance-to-pull-low circuit, 22
Variable-resistance-to-push-up circuit, 22
Wheatstone Bridge, 20
World War I, 7
World War II, 7